UPWELLING

MECHANISMS, ECOLOGICAL EFFECTS AND THREATS TO BIODIVERSITY

Oceanography and Ocean Engineering

Additional books in this series can be found on Nova's website
under the Series tab.

Additional e-books in this series can be found on Nova's website
under the e-book tab.

OCEANOGRAPHY AND OCEAN ENGINEERING

UPWELLING

MECHANISMS, ECOLOGICAL EFFECTS AND THREATS TO BIODIVERSITY

WILLIAMS E. FISCHER

AND

ADAMS B. GREEN

EDITORS

New York

Library of Congress Cataloging-in-Publication Data

Upwelling : mechanisms, ecological effects and threats to biodiversity / William E. Fischer and Adams B. Green.
    pages cm.
  Includes index.
  ISBN 978-1-62948-174-6 (soft cover)
  1. Upwelling (Oceanography) 2. Ocean circulation. I. Fischer, William E.
  GC228.5.U68 2013
  551.46'2--dc23
               2013035646

Published by Nova Science Publishers, Inc. † New York

CONTENTS

PREFACE

An upwelling is a rise of deep sea waters to the surface. A minimum of four types of upwelling have been identified: coastal upwelling; large-scale wind-induced upwelling in the open ocean; upwelling related to tropical cyclones; and upwelling related to topography.

In this book, the authors present current research in the study of the mechanisms, ecological effects and threats to biodiversity due to upwelling. Topics discussed in this compilation include the ecological effects of coastal upwelling in Cabo Frio, Rio de Janeiro, Brazil; Namibian upwelling and its effects on macrozoobenthic diversity; mechanisms of impact on chlorophyll and upwelling in the Northern Black Sea; distribution, mechanisms, and biologic and climatic significance in upwelling and downwelling.

Chapter 1 – In this chapter authors are going to discuss the upwelling of the Cabo Frio region, on the northeast coast of the State of Rio de Janeiro, Brazil, and its ecological effects. This region's main oceanographic characteristic is the occurrence of a coastal upwelling, where cold waters from the South Atlantic Central Water (SACW) emerge onto the continental shelf 24 nautical miles from the coastline (42°30'W – 42°06'W), where subtropical waters (12-15°C and salinity from 35.1 to 35.5‰) constitute the main water mass in the upwelling process. Many studies indicate that SACW is the origin of the cold waters emerging near this coast; others suggest that variations are related to marine topography. The existence of an upwelling at this point of the Brazilian coast is the result of several factors: a) the brusque change in direction of the coastline, passing from north-south to east-west; b) the presence of a structural elevation on the continental shelf; c) the seasonal displacement of the axis of the Brazilian Current, which is deflected offshore during the summer. The intermittent emergence of cold water to the surface is controlled by the wind regime, an essential factor in the variations of intensity of this upwelling.

The ecological effects of the upwelling are easily observed in the region. In Brazil, the coastline is usually quite humid (over 1500 mm/year), originally being covered by rainforest, whereas the interior of the country is drier, especially the Northeastern region, being characterized by a semi-arid climate with a vegetation formation called "caatinga" dominated by xeromorphic forests. The Cabo Frio region, considered a phytogeographical enclave, presents ecological peculiarities, with a drier climate than the rest of the coast of Rio de Janeiro State. This enclave may be explained by the local semi-arid climate (about 800mm/year), caused mainly by the presence of the upwelling. Numerical analyses of climate data will be presented showing that Cabo Frio rainfall is significantly lower than in surrounding areas, more closely resembling the "caatinga" region than its surrounding coast, proving the existence of the climate enclave due to the upwelling. Studies on the current and past local vegetation, rich in biodiversity and endemic species, will also be presented. Paleoenvironmental studies (phytolith analyses) do not indicate major changes in vegetation cover from 13,000 years cal BP, always consisting of dry forests, with relatively drier or wetter periods, which correspond respectively to periods of higher or lower intensity of the upwelling.

Chapter 2 – The Namibian sector of the southern Atlantic shelf is part of the Benguela upwelling system and among the most productive marine ecosystems on our planet. The primary production is significantly enhanced by upwelling of nutrient-rich deep water further charged by nutrients recycled from shelf sediments. Subsurface shelf waters of the Benguela system are characteristically low in oxygen content and periodically anoxic. Free hydrogen sulphide in bottom shelf water has frequently been observed during austral spring and summer and occasional upwelling of sulfidic water at the coast has disastrous impacts on higher life forms. The sedimentary organic carbon load is high compared to oxygen availability resulting in deficiency of dissolved oxygen in bottom waters. These harsh environmental conditions normally hamper the development of higher benthic life. Macrozoobenthos were investigated in this area during the years 2004, 2008 and 2011 at 52 stations between 25 and 2513 m water depths. Altogether 134 grab samples were analyzed. Surprisingly, at all stations and all times macrozoobenthos abundance could be observed, even in oxygen minimum zones (OMZ) (bottom waters with oxygen contents below 0.5 and 0.1 ml/l respectively). Although the biodiversity decreased in the center of the upwelling cells (at water depths of around 50 to 400 m), a high number of well-adapted species remained to form typical benthic communities. Particularly at the fringes of the OMZ biomass an abundance of these species–reduced communities can maintain high values that are comparable to those in oxic environments. As a specific feature of the Benguela System the upwelling cells constitute barriers for species diversity but not for biomass or abundance. This may be an effect of the stability of this system over millions of years, which could have promoted the adaptation of macrozoobenthic communities to extreme environmental conditions.

Chapter 3 – Upwelling manifestations in the Northern Black Sea are described using long-term hydrometeorlogical routine observations and satellite data on the sea surface temperature and chlorophyll-A concentration for the warm season (May through October). It is shown that upwelling in the Northern Black Sea coastal zone is mostly due to rundown or/and Ekman mechanisms and is generated locally under certain wind conditions. During upwelling development, the sea surface cooling exceeds 5°C per 12 hours. Typical duration of an upwelling event is several days. The maximum cooling (up to 22°C) is observed in the coastal regions with narrow shelf. The typical number of upwelling events is one to four per warm season. There is also a rarer type of frontal upwelling which is generated at the edge of meandering Rim current and does not depend on the local wind. So, upwelling in the Northern Black Sea is a sporadic short-term event. However, climatic variations of regional temperature regime depend significantly on the upwelling parameters which are characterized by intense interannual-to-multidecadal variability. In general, long-term weakening of the rundown winds leads to positive trend of surface temperature in the warm season which reaches about 3°C per 50 yrs in some subregions. Upwelling events are accompanied by significant changes of chlorophyll-A concentration in the surface layer of the Northern Black Sea. Lead-lag relation between variations of the sea surface temperature and chlorophyll-A concentration varies from one sub-region to another and crucially depends on the vertical structure of chlorophyll-A concentration.

Chapter 4 – An upwelling is a rise of deep sea waters to the surface. A minimum of four types of upwelling have been identified: (1) coastal upwelling; (2) large-scale wind-induced upwelling in the open ocean; (3) upwelling related to tropical cyclones; and (4) upwelling related to topography. The primary zones of upwelling are located off the eastern boundaries of oceans. A downwelling is a process in the opposite direction. Basic areas of downwelling include the coastal waters of Antarctica (basically, the Weddell Sea) and the North Atlantic (predominately off the Greenland coast). The area of downwelling zones in the world's oceans is much smaller than that of upwelling zones, as the downwelling rates are always 2–3 times larger than upwelling rates.

Upwelling and downwelling are of importance for two kinds of human activity: (1) fisheries and (2) recreational activities. In addition, they play major ecological and climate-forcing roles.

In: Upwelling   ISBN: 978-1-62948-174-6
Editors: W. E. Fischer and A. B. Green © 2013 Nova Science Publishers, Inc.

Chapter 1

ECOLOGICAL EFFECTS OF A COASTAL UPWELLING IN CABO FRIO, RIO DE JANEIRO, BRAZIL

Heloisa Helena Gomes Coe
Department of Geography, Universidade do
Estado do Rio de Janeiro (UERJ), Brazil

ABSTRACT

In this chapter we are going to discuss the upwelling of the Cabo Frio region, on the northeast coast of the State of Rio de Janeiro, Brazil, and its ecological effects. This region's main oceanographic characteristic is the occurrence of a coastal upwelling, where cold waters from the South Atlantic Central Water (SACW) emerge onto the continental shelf 24 nautical miles from the coastline (42°30'W – 42°06'W), where subtropical waters (12-15°C and salinity from 35.1 to 35.5‰) constitute the main water mass in the upwelling process. Many studies indicate that SACW is the origin of the cold waters emerging near this coast; others suggest that variations are related to marine topography.

The existence of an upwelling at this point of the Brazilian coast is the result of several factors: a) the brusque change in direction of the coastline, passing from north-south to east-west; b) the presence of a structural elevation on the continental shelf; c) the seasonal displacement of the axis of the Brazilian Current, which is deflected offshore during the summer.

The intermittent emergence of cold water to the surface is controlled by the wind regime, an essential factor in the variations of intensity of this upwelling.

The ecological effects of the upwelling are easily observed in the region. In Brazil, the coastline is usually quite humid (over 1500 mm/year), originally being covered by rainforest, whereas the interior of the country is drier, especially the Northeastern region, being characterized by a semi-arid climate with a vegetation formation called "caatinga" dominated by xeromorphic forests.

The Cabo Frio region, considered a phytogeographical enclave, presents ecological peculiarities, with a drier climate than the rest of the coast of Rio de Janeiro State. This enclave may be explained by the local semi-arid climate (about 800mm/year), caused mainly by the presence of the upwelling.

Numerical analyses of climate data will be presented showing that Cabo Frio rainfall is significantly lower than in surrounding areas, more closely resembling the "caatinga" region than its surrounding coast, proving the existence of the climate enclave due to the upwelling.

Studies on the current and past local vegetation, rich in biodiversity and endemic species, will also be presented. Paleoenvironmental studies (phytolith analyses) do not indicate major changes in vegetation cover from 13,000 years cal BP, always consisting of dry forests, with relatively drier or wetter periods, which correspond respectively to periods of higher or lower intensity of the upwelling.

Keywords: Cabo Frio, upwelling, semi-arid climate, phytogeographical enclave

INTRODUCTION

The coastal upwelling systems are characterized by the rising of cold water from deeper levels, making nutrients available in the euphotic layer for the enhancement of phytoplankton production and growth. Under these conditions, the primary productivity is quite high, consequently benefitting the entire food chain, which makes these areas important for fishing. Upwelling is most common at the western edges of continents, but on the Brazilian coast it occurs at a number of points, the most intense system being located in the region of Cabo Frio (more precisely in Arraial do Cabo), in the east of the State of Rio de Janeiro (23°S, 42°W) (Figure 1). Studies on the theme date from the 1950s (Allard, 1955) and since then many researchers have been working in this area, for example, Emilson (1961), Moreira da Silva and

Rodrigues (1966, 1973), Moreira da Silva (1977), Signorini (1978) Valentin et al. (1987), Palacios (1993) and Torres Jr. (1995), among others.

The basic knowledge of this system was reviewed in accordance with concepts of biophysical interactions. The high frequency and amplitude of the prevailing winds are the main factor promoting the rise of South Atlantic Central Water (SACW), but meanders and eddies in the Brazil Current (BC) as well as local topography and coast line are also important. Upwelling events are common during the spring and summer seasons. Primary biomass is exported by virtue of the circulation of water and is also controlled by rapid zooplankton predation. Shoreline irregularities define the embayment formation of the Marine Extractive Reserve of Arraial do Cabo, making it an area with evident differences in the intensity of upwellings that harbors high species diversity.

Consequently, on a small spatial scale there are environments with tropical and subtropical features, a point to be explored as a particularity of this ecosystem (Coelho Souza et al., 2012).

The coastal region of Cabo Frio ("Cold Cape"), received its name from the earliest Portuguese navigators to reach there, this being the first indication of the existence of a thermal anomaly of ocean surface waters.

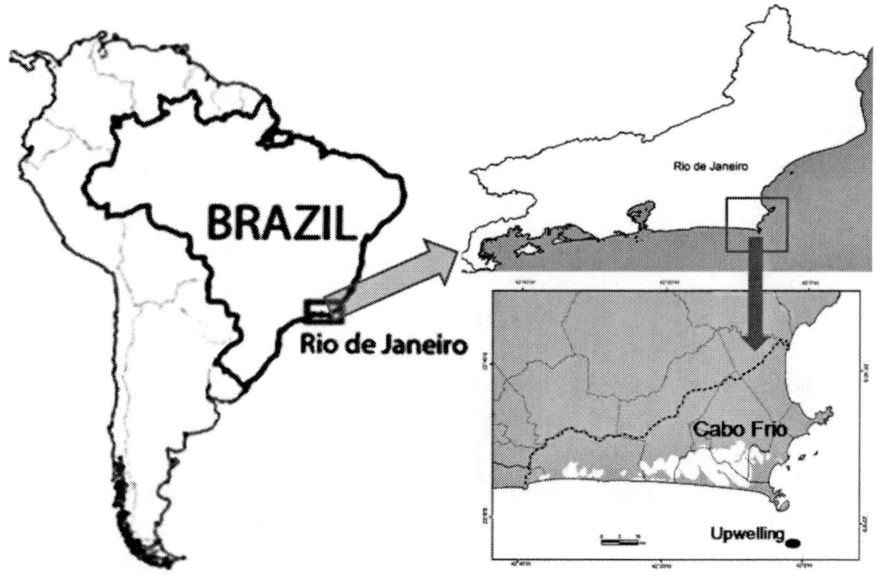

Source: Modified from Google, 2013.

Figure 1. Location of the study area.

Actually, the main oceanographic feature of this region is the occurrence of a coastal upwelling, where cold waters rich in nutrients from the SACW emerge on the continental shelf. Under the influence of the prevailing E-NE winds and the Coriolis deviation, the surface water flows towards the open sea and is replaced by deeper water from approximately 300 m. By penetrating the euphotic layer, this deep cold water (SACW) influences the structure of this tropical coastal ecosystem (Valentin and Coutinho, 1990).

The Cabo Frio region is considered to be a phytogeographical enclave (Ab' Saber, 2003). While rainforest and semi-deciduous forest are the natural vegetation dominating the State of Rio de Janeiro, this region is characterized by xeric formations such as dry forest, xeric shrubland with Cactaceae, also called "caatinga" from comparison with the shrubland typical of the semi-arid northeastern part of Brazil (Mooney et al., 1995), "restinga" shrubland and "restinga" forest (Araujo, 1997). Its vegetation is adapted to the geomorphic diversity of the area and to the local dry conditions, principally due to the coastal upwelling, which leads to a decrease in precipitation and, hence, to the establishment of a semi-arid microclimate (with a mean annual rainfall of 850 mm); while the regional climate, not affected by the upwelling, is tropical humid (with a mean annual rainfall of 1500 mm) (Barbiere, 1984).

The aim of this chapter is to present the upwelling of the Cabo Frio region and its ecological effects, discussing whether or not the region is an enclave. Multivariate analyses of similarity of the current climate variables were performed, comparing Cabo Frio with its humid surroundings and with the "caatinga" of Northeastern Brazil. Furthermore, we seek to infer possible paleoclimatic changes in the region, presenting studies on reconstitutions of vegetation and the intensity of upwelling throughout the Quaternary.

STUDY AREA

Cabo Frio is a cape located in the eastern part of the State of Rio de Janeiro. It is a remarkable point on the Brazilian coast, where its north-south orientation passes sharply to East-West. This reversal is accompanied by a significant change in the profile of the Continental Shelf, taking the 100m isobath very close to shore and lowering most of the shelf to depths of between 100 and 200 m. The contributions of continental water are reduced to two rivers north of Cabo Frio, the Sao Joao and the Paraíba, the more important, which is about 80 nautical miles from the cape.

The area, also known as "Regiao dos Lagos" (The Lakes Region), due to the presence of several coastal lagoons, covers the counties of Arraial do Cabo, Buzios, Cabo Frio, Iguaba, Sao Pedro da Aldeia and Araruama, with an area of approximately 1,500 km^2. It is located between the coordinates 22°30' - 23°00'S and 41°52' - 42°19'W (Figure 1). Despite being located in a humid tropical zone, the region has a semi-arid microclimate, caused by the intermittency of the coastal upwelling.

These particular climatic conditions play a major role in the physical-chemical characteristics of the lagoons and in the control of their sedimentation. Indeed, these lagoons are hypersaline with organo-carbonate sedimentation to the east of the upwelling and hyposaline with organo-clastic sedimentation to the west. Currently, the intensity of the upwelling is controlled by the regime of trade winds of the tropical South Atlantic. This wind regime varies seasonally and inter-annually, especially in response to changes in surface temperatures in the tropical Atlantic, but it also responds to perturbations of the atmospheric circulation, such as those caused by El Niño events. Indeed, during a strong El Niño event, the strengthening of the tropical jet causes a blockage of polar advection south of the region of Cabo Frio, generating permanent winds from the Northeast. This results in an intensification of the coastal upwelling and semi-arid climatic conditions in the lagoon system.

The Cabo Frio region covers a complex geological and geomorphologic setting. Elevation does not exceed 300 m a.s.l. and most of the coastline has been shaped by sea level changes during the Quaternary (Ortega, 1996). The lithology consists mainly of gneiss, sand and clay materials with most soils being poorly developed.

The morphological, chemical and mineralogical properties of the soils in the region suggest a particular pedogenic system, which may represent pedoenvironments, which were formerly broader but nowadays are more isolated being maintained through the regional morphoclimatic peculiarities (Ibraimo et al., 2004).

OCEANOGRAPHIC FEATURES

The upwelling phenomenon is characterized by an upward movement whereby water from the sub-surface layers of the ocean, which are generally cooler, is brought to the surface causing lower temperatures in relation to the average for their respective latitudes (Oda, 1997).

There are several possible causative mechanisms, such as: (1) deep currents meeting an obstacle, for example, an oceanic ridge; (2) divergence between surface currents, for instance waters from north and south of the equator; or (3) removal of water near the coast by wind action. This latter mechanism is the most frequent and causes upwellings known as "coastal" (Braga, 2001), such as the one in Cabo Frio. In coastal upwelling regions, stratification stability is a function of temperature and water currents. The mixing of superficial and deep waters occurs due to the combination of vertical distribution velocity and water density (Allen, 1980; Winant, 1980).

Coastal upwellings are very common all over the world. In Brazil, such areas are observed in seven regions along the southeastern/southern coast (Vitoria, Sao Tome, Cabo Frio, Sao Sebastiao, Santa Catarina, Santa Marta and Rio Grande do Sul). The upwelling intensity is highest in the Cabo Frio region (Rio de Janeiro state), with an extension of 150-400 Km from the city of Arraial do Cabo, contrasting with an extension of 70 Km on the Santa Catarina coast (Kampel et al., 1997).

In the region of Cabo Frio, 24 nautical miles from the coastline ($42°30'W - 42°06'W$) (Figure 1), the subtropical waters (12-15°C e $35.1-35.5^0/_{00}$) constitute the main water mass of the upwelling process (Mascarenhas et al., 1971).

Warm, (>25°C) salty (>36.5‰) water occupies the surface level of the South American Tropical Zone, whose characteristics are a result of intense radiation, excessive evaporation compared to precipitation and weak transportation through the tropical thermocline; providing a maximum salinity layer between the surface and 100 m in depth.

A part of these waters is pushed to the south by the Brazil Current (BC) and contributes, by mixing, to the formation of a mass called "Tropical South Atlantic Water", with temperature and salinity above 20° and 36‰ respectively (Valentin, 1984).

The increase in density through a progressive loss of heat to the atmosphere along its displacement to the south causes the descent in depth of this tropical water. Through this cooling mechanism, the tropical water mixes with the cooler and less salty waters from the south and constitutes the subtropical water (Thomsen, 1962 cited by Valentin, 1984), which is part of the South Atlantic Central Water (SACW) which plunges under BC at the level of sub-tropical convergence (latitude 30-40° S).

Consolidation of the SACW occurs in the Sub-tropical Convergence Region (33-38°S), in the confluence zone of the Falklands and Brazil Currents (Stramma and Peterson, 1990).

SACW is transported by the South Atlantic Current to the African coast, where it joins the Benguela and Agulhas Currents. SACW is also important in the Benguela and Canaries upwelling systems (Pelegri et al., 2005) and turns back towards the South American coast around 16-20°S, flowing northwards with the North Brazil Undercurrent and southwards with the Brazil Current (Stramma and England, 1999).

In Cabo Frio, SACW presents temperatures between 6° and 18°C, salinity between 35 and 36 ‰ and is located generally between 150 and 500 m depth. It rises at a speed of 7×10^{-3} cm/s, when NE winds predominate for periods exceeding 24 hours (André, 1990). Rich in nutrients, SACW significantly increases the nitrate content from 1μM to 10μM and the chlorophyll rates from 0.2-0.5 mg/L to 5μg/L (Valentin, 1994). The increase of fishing activity in the region is associated with this fertilization of waters and to the thermal stratification of the water column. The increasing of stratification causes the formation of a thermocline, the zone of maximum temperature variation, separating surface (warm) and deep (cold) waters. Through the action of NE winds, BC waters are moved away from the coast and replaced by deep cold waters of SACW. The thermocline that exists between them, boosted by the upward movement of cold water, promotes the development of phytoplankton when it reaches the euphotic layer.

With the persistence of NE winds, the thermocline reaches the surface rich in plankton, followed by SACW which then covers the entire water column. This process is usually complete after 3 to 4 days, signaling the start of the period of maximum intensity of the upwelling (Valentin, 1994). During the summer, the internal shelf is occupied by the Coastal Water (CW) which mixes with Tropical Water (TW) further offshore. There is a strong vertical stratification of temperature and salinity due to the SACW upwelling from the deep, through the action of the E-NE winds. In the winter, SACW returns to the deep and the internal shelf is dominated by CW, while TW occupies the external shelf, the vertical stratification of temperature and salinity disappearing (Braga, 2001).

Although the complete mechanism of the upwelling is not fully understood, its relationship with the wind is clear, being based on Ekman's theory, where the average ocean transportation takes place 90° to the left of the wind in the southern hemisphere. On the southeast coast of Brazil, the NE winds, coming from the Semi-Permanent South Atlantic anticyclone, predominate and occur throughout the year, being interrupted in the passage of frontal systems when the direction becomes SW (Torres Jr, 1995; Dourado and Oliveira, 2001).

Thus, surface water tends to move offshore enabling the rise of underlying water, in this case, water from the South Atlantic Center (SAC). The temperature indicative of upwelling is therefore 18°C, a higher thermal index than SACW. Thus, it can be assumed that, in the months that SACW is shallower, less wind power is necessary to bring it to the surface.

Many studies have shown that SACW is the origin of the cold waters that emerge near the coast of Cabo Frio (Miranda, 1985; Valentin et al., 1987, Campos et al., 1995). Other studies suggest that the variation of upwelling along the coast is related to marine topography. Other factors might also be important, including: (1) seasonal variations of the BC position (Stramma and Peterson, 1990); (2) the geostrophic balance determining the vertical movements of the SACW (Signorini, 1978); (3) coastal water suction due to BC eddies, and (4) the introduction of SACW onto the continental shelf induced by meanders of the BC (Campos et al., 2000). The change in the course of the coastline from north-south to east-west defines the Cabo Frio region and this configuration brings continent and continental shelf together. Both coastline geometry and ocean floor topography influence the magnitude of the upwelling (Rodrigues and Lorenzetti, 2001). The topography of the ocean bed also increases the energy and activity of meso-scale eddies (Calado et al., 2010). According to Barbosa (2003), the topographic factors are essential to explain the geographical location of the upwelling, but they do not explain the intermittent risings of cold water to the surface, which are controlled by the wind regime, the essential factor in the variations of intensity of upwelling.

The ascent of cold water takes place as follows: in the summer, warm surface waters of the BC, which skirt the coast towards the south, are deviated to the east. The warm surface waters are carried away from the coast, which causes a rising of cooler and denser deep water from the continental slope over the continental shelf up to a depth of about 50 m, without, however, emerging. This vertical pumping causes a zonal pressure gradient associated with a geostrophic current along the coast (Valentin, 1994).

This wind system can be disturbed by strong "El Niño" events, which cause great changes in the atmospheric circulation in South America. According to Kousky et al. (1984), these anomalies are related to two main processes: (1) the westward displacement of the Equatorial Convection Zone usually centered over the Amazon, which will be positioned over the Eastern Pacific, (2) the reinforcement of the subtropical jet, blocking the polar advections. The blocking zone goes from northern Peru to southern Brazil, passing through Bolivia, the north of Chile and Argentina.

This blockage of polar advections causes intense rainfall in southern Brazil, Paraguay and northern Argentina and a deficit of precipitation in Northeast Brazil. Furthermore, the winds of the southern sector, linked to the increase of the cold fronts, are also blocked in southern Brazil. The position of the blocking zone may vary depending on the characteristics of the "El Niño" event. In the region of Cabo Frio two situations may occur: (1) the blocking zone is located south of Cabo Frio. In this case, the winds of the southern sector do not affect the region. On the other hand, NE sector winds are practically permanent and intensity of upwelling will be enhanced; (2) the blocking zone is located north of Cabo Frio. In this case, NE sector winds do not affect the region and, consequently, the resurgence is attenuated and sometimes may even disappear (Valentin, 1994).

According to Barbosa (2003), negative temperature anomalies of the sea surface caused by the direction of the winds in coastal upwelling are present on the southeast Brazilian continental shelf, especially during the summer. When NE winds persist for several days, a strong upwelling may occur, with the decrease of surface temperature to 15°C (or less) near the coast of Cabo Frio. These temperatures are around 10°C cooler than those of external or medium shelf waters. The surface layers of the region of the shelf break are usually occupied by the waters of the BC, with high temperatures and salinity. The BC can reach temperatures of 25°C to 27°C in summer and 22°C to 24°C during the winter. The salinity generally ranges between $36.5^0/_{00}$ and $37^0/_{00}$ (Castro and Miranda, 1998). When cold fronts occur, surface winds rotate, causing a strong windstorm from the southern quadrant, thereby inhibiting the upwelling (Stech and Lorenzzetti, 1992).

Valentin (1994) summarizes the two phases of the Cabo Frio upwelling as follows: the first phase is seasonal; SACW goes beyond the continental slope and invades the bottom of the shelf, where it stays from September to April. In this period BC is away from the coast. In the second phase, the direct action of the local winds from the eastern sector, with high frequency and intensity during spring and summer months, promotes, at this time, the rise of SACW to the surface. The upwelling cycle is interrupted more frequently during the winter (Valentin, 1994). When the SW winds prevail, in the passage of cold fronts, they provide the piling of surface waters on the shore, making the SACW return to the deep. In this case, a reverse phenomenon of upwelling takes place; downwelling. The intensification of the Cabo Frio upwelling leads to a reduction in precipitation and, therefore, an increase in the dryness, evaporation and salinity of the lagoons, likewise, a weakening leads to the opposite effect.

Certainly, similar variations are produced on an annual scale with significant impact on the lagoon environment, especially in regard to the carbonate sedimentation in smaller and isolated lagoons, controlled by the local microclimate.

EFFECTS OF UPWELLING ON CABO FRIO'S CLIMATE

The negative anomalies of sea surface temperature (SST) increase the thermal contrast between ocean and continent, causing an intensification of breeze circulation. In Cabo Frio, wind data time series show signs of the sea breeze being more pronounced during the summer, indicating that the upwelling must intensify it (Miranda, 1985). Numerical modeling studies of the local circulation of Cabo Frio also show greater intensity of breeze circulation during the period from October to March and weaker intensity between April and September (Franchito et al., 1998; Oda, 1997).

The local atmospheric circulation in Cabo Frio is greatly affected by the sea/land breeze circulation. However, as the local SST site is dominated by seasonal variation, in periods of lower SST there is greater diurnal variation in the magnitude of the wind and less variation in its direction. During the months of highest SST, diurnal variation in direction is greater because of the alternation between sea breeze and land breeze. Considering the local SST as a function of the presence or absence of coastal upwelling, which is driven by large scale circulation, these variations are also present in wind series. It can be concluded, therefore, that the zonal component seems to depend on the diurnal continent heating, while the meridional component variations follow large scale circulation, showing longer period variations (Oda, 1997).

The region of Cabo Frio, in the Zone of Influence of Upwelling (Barbiere, 1999), is considered an enclave, by presenting a microclimate with low rainfall, distinct from the tropical humid climate dominant along the southeastern coast of Brazil. The local climate is classified as semi-arid hot, a variation of the Köppen Bsh (Barbiere, 1975). It is characterized by low rainfall (on average 770-854 mm / year) and an evaporation rate between 1200 and 1400 mm / year (Barbiere, 1984), especially in the summer period. The average temperature is slightly above 21°C from June to September and varies between 23 and 25°C from November to April, not showing, therefore, a large annual temperature range. Insolation varies between 200 and 240 h/month (Barbiere, 1984), with the exception of the period between September and November when it ranges from 150 to 190 h/month (Table 1).

Table 1. Cabo Frio - Month Norms from 1961 to 1990

CABO FRIO (Álcalis)/RJ Climatolological Station					
Latitude: 22°59' S Longitude: 42°02' W Elevation: 7.00 m					
MONTH	EVAPORATION (mm)	INSOLATION (hs)	PRECIPITATION (mm)	AVERAGE COMPENSATED TEMPERATURE[*] (°C)	RELATIVE HUMIDITY (%)
JAN	80.9	239.8	78.1	25.1	82
FEB	78.5	235.2	44.1	25.4	82
MAR	77.4	227.8	52.8	25.4	82
APR	78.1	197.7	78.3	24.3	80
MAY	71.2	214.3	69.1	22.8	81
JUN	67.5	201.1	43.9	21.6	81
JUL	78.3	218.5	44.7	21.3	80
AUG	79.8	203.7	36.1	21.2	81
SEP	83.3	156.2	61.0	21.3	81
OCT	78.7	179.1	80.7	22.2	82
NOV	79.8	189.6	81.0	23.3	82
DEC	78.6	201.6	101.1	24.5	82

Source: INMET, 2005.

[*]Average Compensated Temperature: the average of three measurements (9h, 15h and 21h, for example) plus the maximum and minimum temperatures of the day.

The region is characterized by the existence of two well-defined seasons: a wet summer season and a dry winter (Nimer, 1989). Summer is characterized by the predominance of NE winds, while winter is marked by discontinuous periods of S-SW winds, linked to the passage of frontal systems from middle latitudes.

The explanations for the occurrence of climatic differences in a short space are based on several factors, due to the geological, geomorphological, oceanographic and climatic peculiarities of the region of Cabo Frio (Figure 2).

Many of these peculiarities originate in the palaeoevolutive history of the region, with climate fluctuations (alternation of warmer/ more humid periods and colder/ drier ones) and consequent changes in relative sea level during the Quaternary; with marine regressions during glacial periods and transgressions, mainly in Holocene, during interglacial periods.

Local geomorphological factors such as the fact that Cabo Frio is a cape, associated with the great distance of the "Serra do Mar" mountain range from the coastline also influence the local climate. This is the point of the State of Rio de Janeiro where the "Serra do Mar" is furthest from the coast, reducing the possibility of orographic rainfall formation.

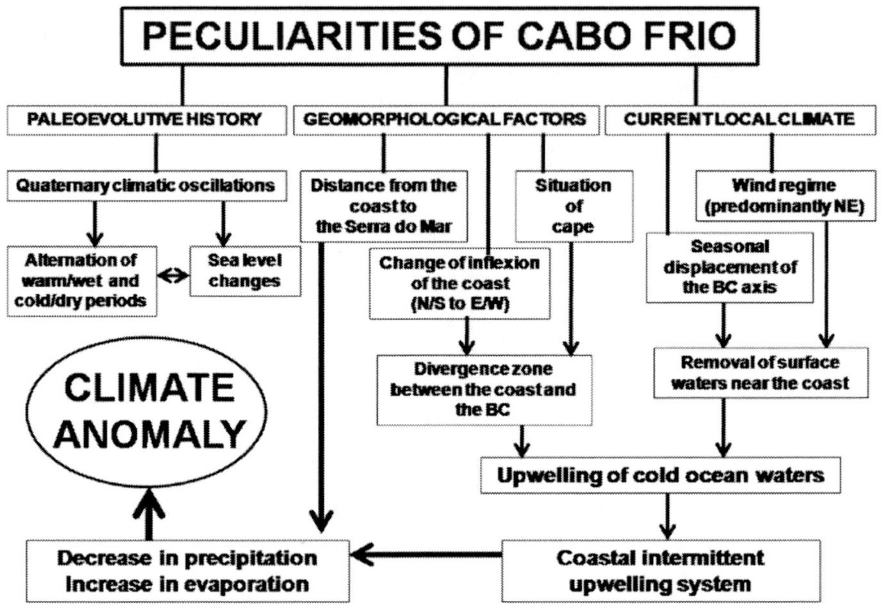

Figure 2. Geological, geomorphologic, oceanographic and climatic peculiarities of the region of Cabo Frio.

The orographic control is indicated by Barbiere (1999) as an important factor of rainfall distribution, whereby rainfall totals show a significant reduction from the top of the "Serra do Mar" towards the coast. Moreover, at this point the continental shelf is less wide and there is a structural high where the Brazilian coast direction changes from north-south to east-west, an inflexion that causes a zone of divergence between the coast and the Brazil Current (BC).

The most important factor for the aridity observed in Cabo Frio seems to be the effect caused by upwelling (Martin and Sugio, 1989; Turcq et al., 1999). The distribution of cold waters in the area of influence of the cape inhibits the formation of cumulus responsible for convective rainfall, thereby decreasing the precipitation rate.

Figure 3 shows a map of isohyets of the eastern sector of the coast of the State of Rio de Janeiro. The stations of Cabo Frio, Iguaba Grande and Saquarema, located in the area of influence of the upwelling, register less than 1,000 mm annual rainfall, while other places, just over 100 km away, can have more than 2,500 mm of annual rainfall.

Modified from http://www.cprm.gov.br/publique/media/Isoietas_Totais_Anuais_1977-2006_2011.

Figure 3. Map of isohyets of the State of Rio de Janeiro, highlighting the 800 mm isohyet and the Cabo Frio region.

The total rainfall and vegetation physiognomy of Cabo Frio led to several hypotheses of similarity between this area and the "caatinga" region of northeastern Brazil (Ab'Saber, 1973, 1977, 2003; Araujo, 2000, Barbiere, 1975, 1984, 1986; Nimer, 1989). In order to more objectively confirm these hypotheses, we performed some statistical analyses comparing climatic variables of precipitation, evaporation, insolation, temperature and relative humidity between the two regions.

To compare the climate of Cabo Frio with its surroundings, we (Coe and Carvalho, 2013a) used rainfall data available from 15 meteorological stations located within 150 km of Cabo Frio (Table 2), covering observation periods longer than 14 uninterrupted years, but in different years (between 1935 and 1993, depending on the season). For the "caatinga" region (34 meteorological stations in the Northeast Brazil region), it was possible to use evaporation data (Eva); insolation (Ins), precipitation (P), average temperature (Tar) and relative humidity (RH) recorded in the same period of 30 years (1961-1990).

Precipitation

Comparing a data series of some meteorological stations in the state of Rio de Janeiro with others in the region of "caatinga" (Northeast Brazil and northern Minas Gerais) (Figures 4 and 5), it can be seen that in Cabo Frio average rainfall is 750 mm/ year (Table 2), while in other areas of the state rainfall is over 1100 mm/ year, reaching almost 2000 mm/ year in areas of higher rainfall (Angra dos Reis). However, several stations of the caatinga present annual totals similar to Cabo Frio, around 800 mm/ year, as observed in Figures 4 and 5.

Insolation

The data show that the total annual hours of sunshine in Cabo Frio are also higher than those of other stations in the state of Rio de Janeiro (about 2500 h/year compared to 1700-2200 h/year), this being, however, average for stations of the "caatinga" (Figure 6). Although there is a great latitudinal difference between the two regions (Figure 7), the similarity of insolation between Cabo Frio, RJ, and other stations of the "caatinga" may be associated with the relative humidity and the frequency of cloud cover.

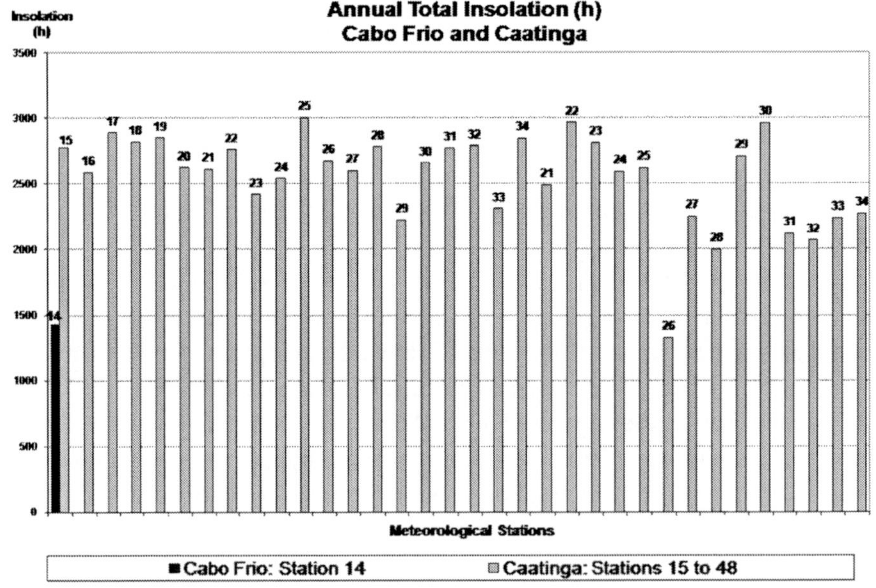

Source: INMET, 2005.

Figure 4. Annual total (mm) of precipitation from meteorological stations around Cabo Frio and "caatinga" region.

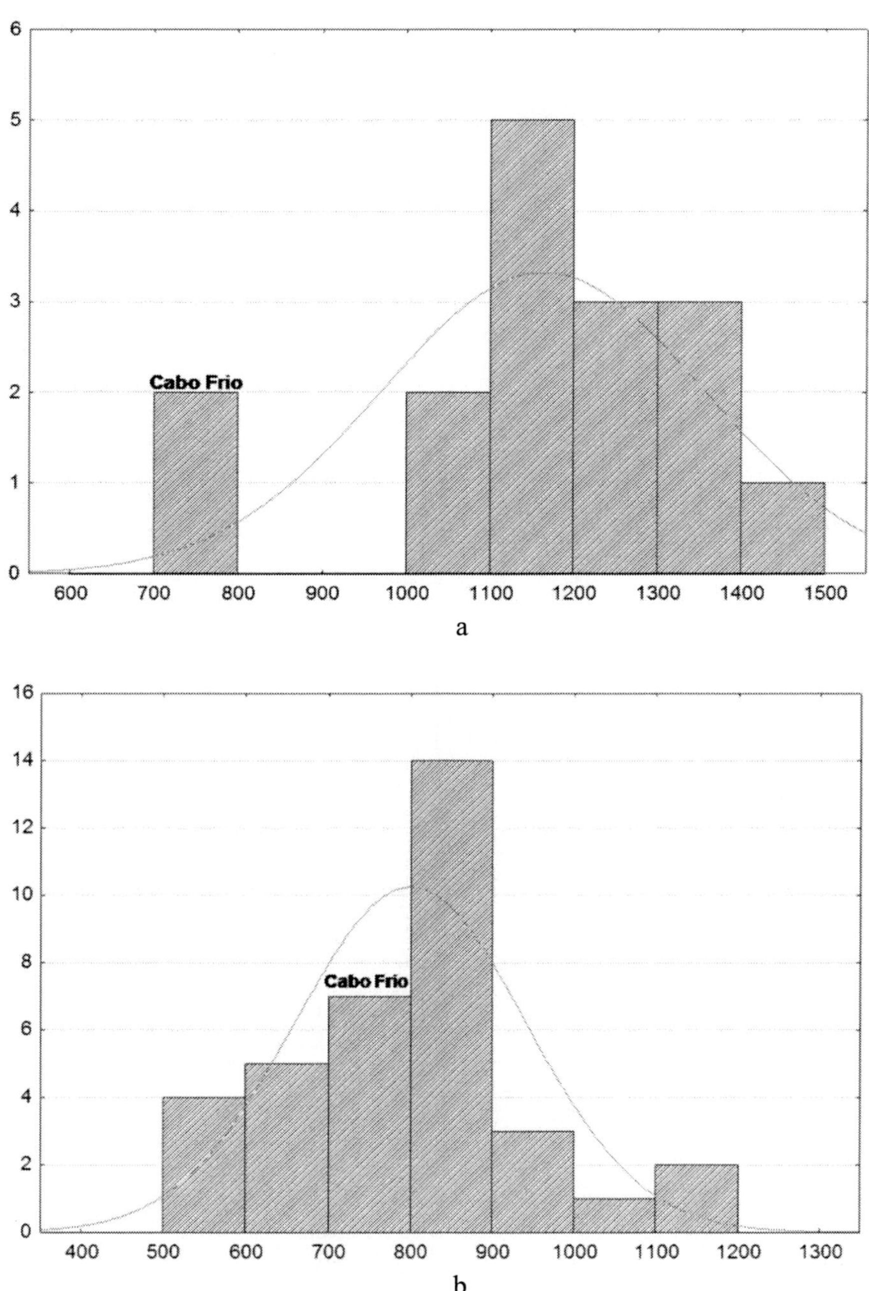

Figure 5. Comparison of the total annual rainfall: a) Cabo Frio and its surroundings; b) Cabo Frio and Caatinga.

Heloisa Helena Gomes Coe

Table 2. Total annual rainfall at some stations in the state of Rio de Janeiro (Cabo Frio and its surroundings) and the area of occurrence of "caatinga" in the Northeastern Brazil Region

STATION	Annual Total (mm)	STATION	Annual Total (mm)
Rio de Janeiro		Caatinga	
1- Cabo Frio	752	25-Vitória da Conquista (BA)	734
2- Maricá	1055	26-Morro do Chapéu (BA)	749
3-Barra de São João	1140	27-Itaberaba (BA)	763
4-Niterói (Horto Florestal)	1141	28-Mossoró (RN)	766
5-Itaboraí	1143	29-Campina Grande (PB)	803
6-Rio Bonito	1150	30-Picos (PI)	812
7-Rio de Janeiro	1173	31-Cariranha (BA)	814
8-Niterói (Ilha do Modesto)	1201	32-Crateús (CE)	826
9-Palmital (Saquarema)	1260	33-Bom Jesus da Lapa (BA)	830
10-Manoel Ribeiro (Maricá)	1268	34-Monteiro (PB)	839
11-Rio Mole (Saquarema)	1303	35-Araçuaí (MG)	841
12-Tanguá	1371	36-Senhor do Bonfim (BA)	851
13-Sambaetiba (Itaboraí)	1374	37-Jacobina (BA)	851
14-Juturnaíba (Araruama)	1432	38-Quixeramobim (CE)	858
Caatinga		39-Palmeira dos Índios (AL)	869
15-Cabrobó (PE)	517	40-Garanhuns (PE)	870
16-Paulo Afonso (BA)	583	41-Pedra Azul (MG)	877
17-Paulistana (PI)	597	42-Caetité (BA)	891
18-Macau (RN)	600	43-Apodi (RN)	920
19-Petrolina (PE)	610	44-Tauá (CE)	926
20-Campos Sales (CE)	619	45-Sobral (CE)	960
21-Barra (BA)	661	46-Barbalha (CE)	1001
22-Arco Verde (PE)	694	47-Floriano (PI)	1103
23-Remanso (BA)	696	48-Bom Jesus do Piauí (PI)	1157
24-Florania (RN)	728		

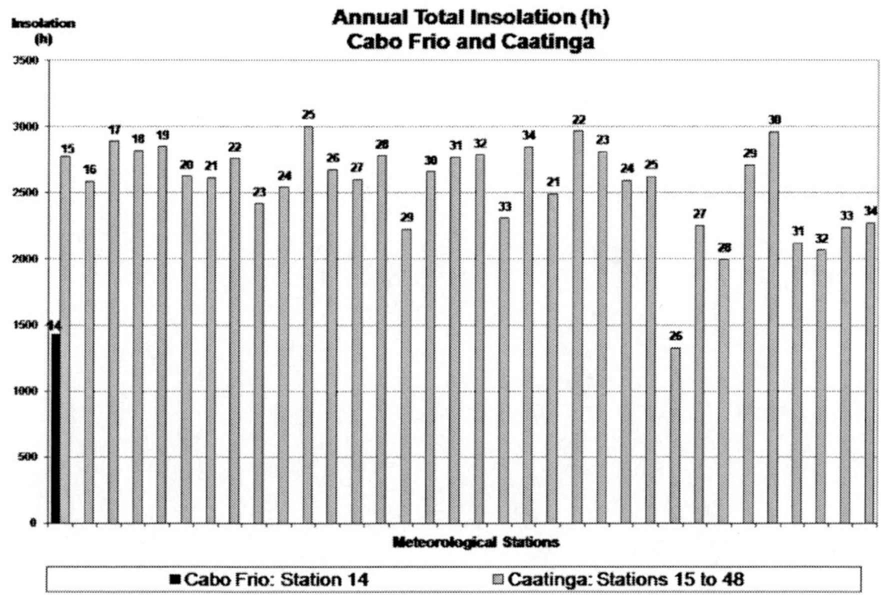

Source: INMET, 2005.

Figure 6. Insolation (total hours/ year) in the meteorological stations of Cabo Frio and in the "caatinga" region.

Average Annual Temperature

The average annual temperature (average of compensated average temperature of 12 months) of Cabo Frio, in the period from 1961 to 1990, was 23.2°C. The differences of temperature between Cabo Frio and the stations of the caatinga area are not significant (Mann-Whitney test, $\alpha/2 = 2.5$), confirming the similarity between both regions.

Multivariate Analyses

Making an analysis by Box plot (Figure 8), it is found that Cabo Frio's precipitation has values considered to be outliers when compared to its surroundings.

However, in comparison with the caatinga, the region of Cabo Frio fits perfectly into the median values, considering precipitation, evaporation, insolation and relative humidity.

Modified from Google Earth, 2012.

Figure 7. Location of meteorological stations in the state of Rio de Janeiro and in the area of occurrence of "caatinga" in Northeastern Brazil.

After an exploratory analysis of the data and comparison of average rainfall, a multivariate grouping analysis with Euclidean distance between the places matrix and amalgamation of groups by the method of minimum variance was done, using the standardized variables. To get around the fact that the local series are temporal heterogeneous, variable standardization and non-parametric tests were preferred.

Cluster analysis confirmed the similarity between Cabo Frio and the "caatinga" region. The most similar places to Cabo Frio are the counties of Palmeira dos Indios, AL (1.25); Araçuai, MG (2.17), Senhor do Bonfim, BA (2.40) and Jacobina, BA (2.88) (Figure 9).

Analyses proved, therefore, the existence of the climatic "enclave" in Cabo Frio. This pluviometric contrast compared to the rest of the State of Rio de Janeiro provides two distinct climates within a reduced distance; with a tropical humid climate dominating the State and a semi-arid climate in Cabo Frio.

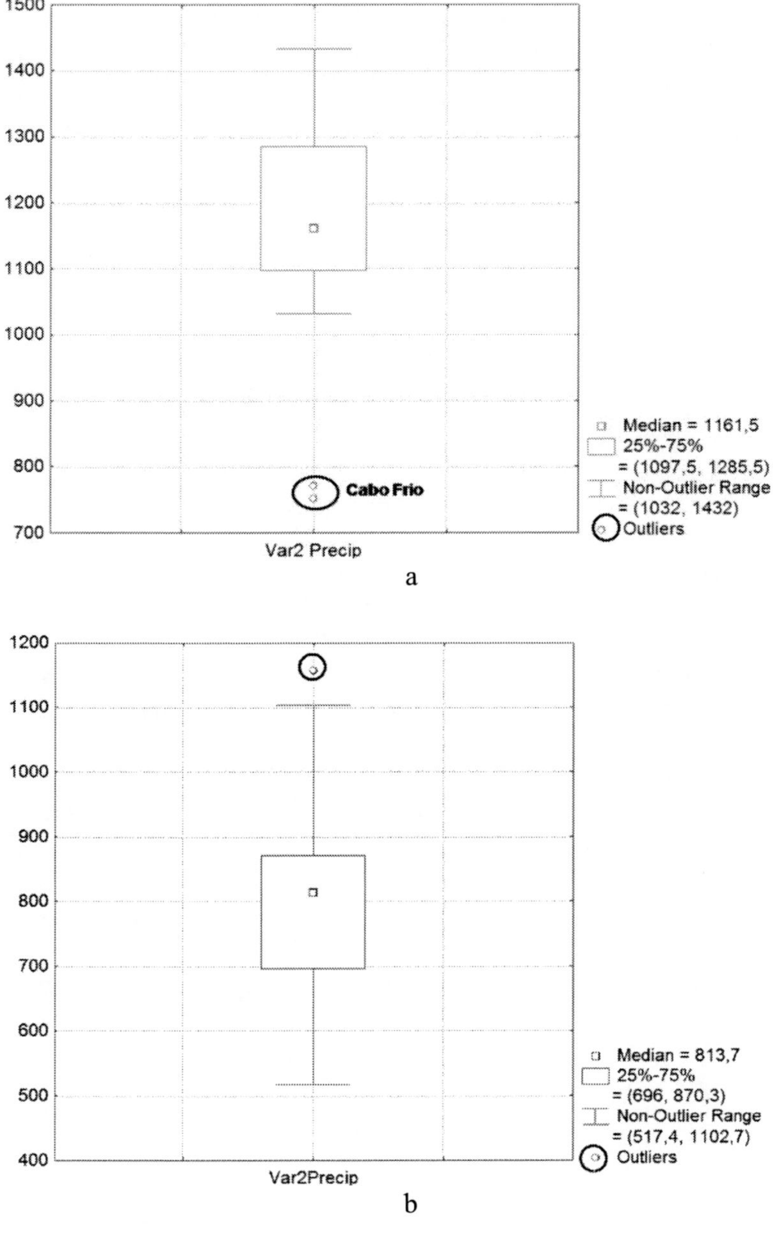

Figure 8. Box plot: a) precipitation of Cabo Frio and its surroundings; b) precipitation, evaporation, insolation and relative humidity of the "caatinga" region and Cabo Frio.

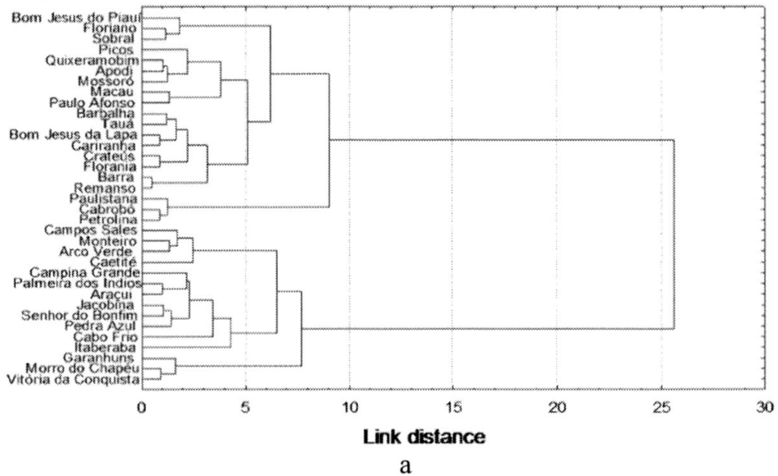

Figure 9. Multivariate analysis (Cluster with distance matrix) comparing meteorological stations of Cabo Frio and 34 stations of the "caatinga" region, using the variables: evaporation, insolation, precipitation and relative humidity.

EFFECTS OF UPWELLING IN VEGETATION

The geoecological peculiarities of Cabo Frio affect various plant formations, with many rare and endemic species. Oceanic-atmospheric physical processes are essential to the peculiar climate that determines the semi-arid vegetation of the Cabo Frio region and which contrasts with that of other parts of the Brazilian coast (Franchito et al., 1998; 2007; 2008). This is an expected pattern since desert or arid adjacent land areas are commonly found in the regions of upwelling systems (Odum, 1983).

The drier climatic condition of the region defines the plant formations that do not present the exuberant aspect that the hillside forests of the State of Rio de Janeiro normally do (Farag, 1999). The forests of this region were classified as a "physiognomic-ecological disjunction of the Northeastern steppe" (Ururahy, 1987) or as "dry forests" (Mooney et al., 1995).

These forests fit perfectly into the definition of dry forest proposed by Mooney et al. (1995) - "in simple terms, they are forests that occur in tropical regions, where there are many months of severe, sometimes absolute drought." Neotropical dry forests are generally less rich in species than rainforests. Thus, what stands out in dry forests is not the diversity of species but their adaptation to hydric stress and disturbances. The Cabo Frio region has 65% of the species indicated by Araujo (2000) as being endemic to the State of Rio de Janeiro, a much higher percentage than any other "restinga" region. That was one of the reasons that led WWF/IUCN to indicate Cabo Frio as a Center of Plant Diversity. Among all physical factors, the climate seems to be the element that has the most influence on the local ecosystems. The remarkable climate action is visible on the dry physiognomy that characterizes the forests of Cabo Frio, especially in places where there is no accumulation of water or protection from the strong sea winds, rich in saltpeter. The wild nature was reported by Ule (1967), commenting on the lack of tropical features in these forests, highlighting the great aridity, despite the lagoons and marshes of the region, which can be observed in the scarcity of mosses that grow on the stems and branches. This lack of epiphytes and mosses demonstrates the influence that low precipitation has on vegetation. Another relevant aspect in a preliminary floristic survey of the flora of Cabo Frio is the absence of species of the Lauraceae family (Araujo and Maciel, 1998). This data is inconsistent with the rest of the forests of Rio de Janeiro, because, according to Lima and Guedes-Bruni (1997) and Marques (1997, cited in Farag, 1999), Lauraceae is one of the 10 families richest in species and, even in the nearby "restingas" seven species can be observed (Araujo and Maciel, 1998).

Photo: Coe, 2007.

Figure 10. The endemic Cactaceae *Pilosocereus ulei*.

Along the entire coast from the island of Cabo Frio to Buzios, a unique floristic composition can be observed in small xeromorphic forests on the cliffs near the sea (Araujo, 1997), whereby the *Pilosocereus ulei* columnar cactus (Figure 10), endemic to the region (Rizzini, 1979), confers a characteristic appearance of arid environments during the driest seasons of the year, similar to the "caatinga". Araujo (1997), based on the comparison between the areas of "restinga" of the coast of Rio de Janeiro, cites Cabo Frio as being the richest in species.

The region also has most of the endemic species of Rio de Janeiro's coastal plain, with 26 of the 36 listed as endemic to the "restingas"; among which 11 endemic species have been identified in dry hills. The natural vegetation is composed of a mosaic of forms that found in low precipitation the most limiting and selective agent (Araujo, 1997), being comprised of (1) tree or shrub "restinga" formation, characterized by medium sized trees and shrubs adapted to dry and nutrient-poor conditions, covering the sand barriers; (2) mangroves and marshes (covering lagoon shores and flood environments); (3) deciduous dry forest composed of dense trees with thin trunks, 3 to 10 m high, on the colluvial-alluvial plain (Ibraimo et al., 2004); (4) rainforest, also called the Atlantic low-montana forest (Rizzini, 1979), at wind-protected or humid sites; (5) xeric shrubland with predominance of Cactaceae (especially the endemic species *Pilosocereus ulei*), with little floristic diversity but adapted to water stress (Araujo,1997), covering the areas most exposed to sea breezes, like the coastal massifs. Pastures and urban areas are today widespread.

Araujo (1997) identified three physiognomic units in the region: (1) coastal plain (beaches, dunes and lowlands, flooded areas, lagoons and alluvial deposits), (2) low hills of Cabo Frio, Buzios and coastal islands; (3) continental hills up to 300 m (Figure 11).

In the areas most exposed to salt spray and sea breezes, the coastal hills are covered with a low forest (3 meters tall on average) composed of dense trees with thin trunks. In places protected from the wind, in moist ravines or in the hills farther from the sea (such as Sierra de Sapiatiba), the vegetation takes on a more robust size, similar to the low-montana Atlantic forest (Rizzini, 1979).

Photos Coe, 2007.

Figure 11. Vegetation in different physiognomic units of Cabo Frio region: a) beaches; b) dunes; c) flooded areas; d) Low hills; e) coastal islands; f) continental hills up to 300 m.

In the colluvial-alluvial plain, the vegetation is characterized as a forest formation with great predominance of deciduous species, taking, during the driest months of the year (June/August), a grayish appearance. The upper stratum presents the canopy with an average height of 10 meters, and may have emergent trees of up to 17 meters high.

The region has several endemic species or those with restricted or disjointed distribution in relation to the "caatinga" area. From a matrix of flora species of Cabo Frio, based on pre-existing flora surveys (Araujo, 1997, 1998, 2000; Farag, 1999; Lima, 2000; Prado, 1991), we identified a number of species presenting such disjointed distribution patterns between Cabo Frio and "caatinga", listed below:

1- *Aechmea lamarchei* Mez (Bromeliaceae); *2- Jacquinia brasiliensis* (Theophastacea); 3- *Adenocalymma comosum* (Cham.) DC. (Bigoniaceae); 4- *Couepia ovalifolia* (Schott) Benth (Chrysobalanaceae); 5- *Croton migrans* Casar (Euphorbiaceae); 6- *Herreria salsaparilha* (Herreriaceae); 7- *Swartzia apetala* (Papilionoidae); 8- *Astronium graveolens* (Anacardeaceae); 9- *Alseis involuta* K. Schum (Rubiaceae); 10- *Bauhinia albicans* (Leg. Caesalpineae); 11- *Brunfelsia latifolia* (Solanaceae); 12- *Caesalpinia ferrea* Mart. ex Tul. (Caesalpinaeceae); 13- *Conchocarpus heterophyllus* (Rutaceae); 14- *Machaerium Albicans* (Papilionoidae); 15- *Opuntia brasiliensis* (Cactaceae); 16- *Oxandra nítida* (Annonaceae); 17- *Pilosocereus ulei* (Cactaceae); 18- *Skytanthus hancorniaefolius* (A.DC.) Miers (Apocynaceae).

CABO FRIO'S UPWELLING AND CLIMATE VARIATIONS DURING THE QUATERNARY

Some hypotheses have been proposed to explain the vegetation dynamics in the region during the Quaternary. Ab'Saber (1977), based on studies of current flora and fauna as well as geomorphological evidence, postulates that, with the environmental changes in the last glaciation, the "caatinga" vegetation came to extend over a large part of Tropical Atlantic Brazil, remaining in some places since then as mini or meso enclaves, where there is predominance of Cactaceae and Bromeliaceae. The evidence therefore suggests that Cabo Frio would have been a paleoclimatic witness to the cold, dry climate of the last glaciation of the Quaternary.

Prado (1991, 2000), through a botanical study, established a model for the expansion of seasonally dry forests in South America which identifies three distribution nuclei: 1) the caatinga of Northeast Brazil, 2) the forest system of the Parana-Paraguay river basin, and 3) the forests of the Sub-Andean southwestern piedmont of Bolivia and northwest Argentina. These three nuclei would be linked by two routes of connection (NW-SE and NE-SW). The NE-SW route would cross over the State of Rio de Janeiro. Lima (2000), through the identification of typical caatinga floristic elements in some parts of the coast of Rio de Janeiro (mainly in Cabo Frio), reinforces the previous model as the large number of species endemic to the dry forests in Rio de Janeiro would be related to their isolation during the Holocene.

Araujo (2000), also through botanical studies, explains the development of the flora of "restingas" along the entire Brazilian coast during periods of marine regressions (during the Pleistocene and Holocene). The difference in floristic composition of current "restingas" could be explained by a compression of this flora during marine transgressions in discontinuous marine terraces.

The Cabo Frio region has also been the object of paleoenvironmental studies (Coe, 2009; Coe et al., 2013b), which were done in order to infer changes in vegetation and, hence, climate during the Quaternary. For this purpose, phytoliths, complemented by isotopic analyzes (δ^{13}C) and lignin phenols were used as proxies. Samples were dated by ^{14}C-AMS. Phytolith indices of tree density (D/P) were calculated and they indicate open vegetation formations, with few trees, with little variation through time.

These phytolith analyses do not indicate a major change in the type of vegetation in the region: since 13,000 years cal BP vegetal cover has always been dominated by xeric forests, suggesting that the local vegetation has not reached the characteristic tree density of the rest of the rainforests of the Rio de Janeiro coast.

The results of the phytolith analyses were correlated with other paleoenvironmental studies on the region.

If we interpret periods of higher tree density as more humid and the lower density ones as drier, we can infer that 13,000 years cal BP, in the late Pleistocene, the climate in the region of Cabo Frio was drier than today.

From 13,000 to 9000 years cal BP, geochemical studies of organic matter from ocean sediments (Andrade, 2008) indicate a period of fast sea level rise. Laslandes et al. (2006), studying coccolithophores compared with lagoon records, identified a weak upwelling period from 12,900 to 11,830 cal years BP, and an intense upwelling phase from 11830 to 8140 cal years BP, when strong El Niño events were recorded (Martin et al., 1988; Turcq et al, 1999).

These results corroborate the ones obtained from phytolith analyzes, indicating, in 8170 years cal BP, a climate similar to that of today.

For the period between 8140 and 7540 years cal BP, Andrade (2008) observed weak or absent upwelling episodes, which corroborates the results of a more humid climate in 7425 years cal BP inferred through phytolith analyses.

Studies of lagoon sediments (Ortega, 1996; Ireland, 1987) indicate that 7,000 years cal BP, the current average sea level was reached. During this period, Andrade (2008) reported a weakening of upwelling, which seems to have extended up to 6000 cal years BP, according to the studies of foraminifera (Oliveira, 2008), corroborating the results of phytolith analyses that infer a more humid climate in 6210 years cal BP.

According to sedimentological studies (Ortega, 1996), sea level continued rising until 6500-5100 cal years BP (maximum of the last transgression). The phytolith results indicate a climate similar to the current 5600 years cal BP.

Oliveira's studies (2008) indicate the beginning of the Holocene climatic optimum and intensification of ENSO at 5,000 cal years BP. Anthracological studies (Scheel-Ybert, 2000) also recorded relatively humid episodes from 5,500 to 4.900/4.500 cal years BP. These studies corroborate the results of a more humid climate in 4500 years cal BP inferred by phytolith analyses.

Studies on the local upwelling (Andrade, 2008; Oliveira, 2008) indicate strong upwelling and El Niño events from 6,000/5,000 to 2,500 cal years BP and a large climate variability (ITCZ, ENSO, NE winds), with frequent but weaker than the previous upwellings from 2,500 cal years BP. After 4520 years cal BP up to 2,760 cal years BP, phytolith analyses do not infer changes in tree density in Cabo Frio's vegetation, which seems to be similar to the current one.

Figure 12 presents the main variations of D/P index during the studied period, inserting a summary of other paleoenvironmental interpretations mentioned in the literature.

Analyzing figure 12, we can observe major trends (periods of higher or lower tree density) in the evolution of the vegetation of the region of Cabo Frio. From 13ka cal BP, there have been no major changes in the vegetation cover of xeric forest. From 13ka cal BP to 6ka cal BP, the tendency is increasing tree density.

This trend is corroborated by records of weaker upwelling events (Andrade, 2008; Oliveira, 2008). From 6ka cal BP to 1.5 ka cal BP, the trend is a decreasing in tree density, corroborated by records of more intense upwelling and El Niño episodes (Andrade, 2008; Oliveira, 2008).

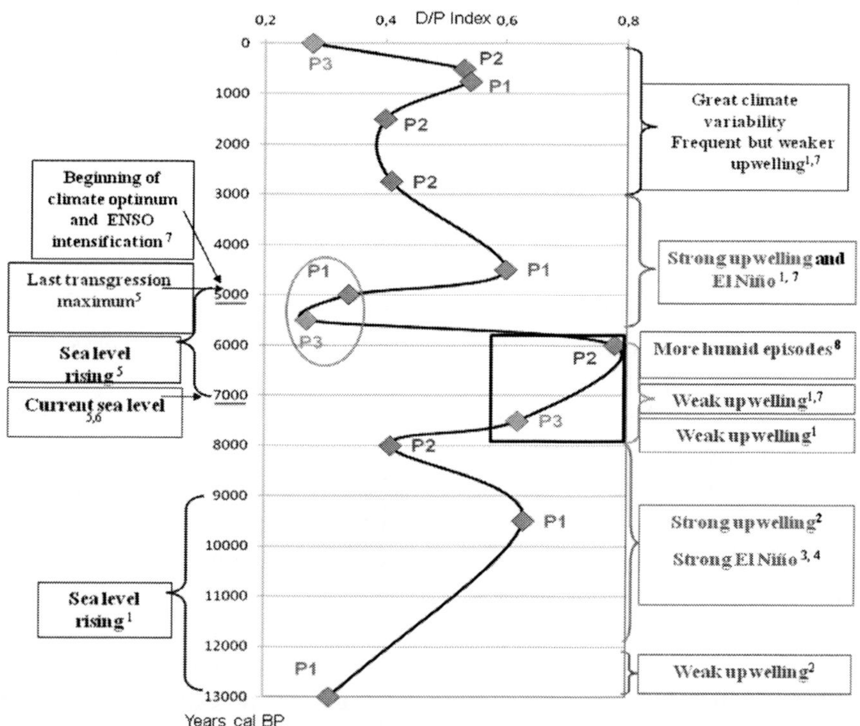

Figure 12. D/P index during the Holocene compared with other palaeoenvironmental studies (identified by reference numbers) in the Cabo Frio region: 1) Andrade, 2008; 2) Laslandes et al., 2006; 3) Martin et al., 1988; 4) Turcq et al., 1999; 5) Ortega, 1996; 6) Ireland, 1987; 7) Oliveira, 2008; 8) Scheel-Ybert, 2000.

During this phase, the period of lowest tree density was between 6 and 5ka cal BP, when we have the lowest D/P index, enrichment in $\delta13C$ and an increased ratio of C/V lignin phenols, characteristic of less dense vegetation. From 1.5 ka cal BP, the phytolith signal no longer records variations and upwelling studies point to a period of great variability.

Variations in sea level recorded at a regional level (eastern coast of Brazil) do not seem to have a direct relationship with changes in the vegetation cover of Cabo Frio. These changes seem to be more influenced by variations in the intensity of upwelling, at a local level.

There appears to be good concordance between the results of upwelling and phytolith studies, despite some differences due mainly to the time scale of the changes that can be registered by upwelling studies and the response of the vegetation to these changes (recorded through phytolith analyses).

Phytolith analyses record variations in vegetation which may occur at local level without necessarily being linked to major climate changes, on a regional scale.

CONCLUSION

In this chapter we presented the coastal upwelling of the region of Cabo Frio, Rio de Janeiro, Brazil, and its ecological effects, discussing whether or not the region is an enclave and if there were variations in the intensity of upwelling during the Quaternary.

Cabo Frio upwelling induces a reduction in precipitation; therefore the region is considered an enclave, with a microclimate distinct from the humid tropical dominant on the Southeast Brazilian coast. The statistical analyses that were performed confirmed the existence of the climatic enclave, which is responsible for ecological peculiarities that affect various plant formations, with many rare and endemic species.

Paleoenvironmental studies indicate that the climatic enclave of the Cabo Frio region has existed since the beginning of the Holocene, alternating relatively more or less wet periods, but always having more arid characteristics than the rest of the Southeast Brazilian coast. Variations in sea level recorded at a regional level (eastern coast of Brazil), showed no direct relation to changes in Cabo Frio's vegetation. These changes seem to be more influenced by variations in the intensity of upwelling, at a local level.

REFERENCES

Ab'Saber, A. N. 1973. A organização natural das paisagens inter e subtropicais brasileiras. *Geomorfologia*, 41: 1-39.

Ab'Saber, A. N. 1977. Espaços ocupados pela expansão dos climas secos na América do Sul por ocasião dos períodos glaciais quaternários. *Paleoclimas*, 3: 1-19.

Ab'Saber, A. N. 2003. Redutos de cactáceas, jardins da natureza. *Scientific American Brasil*, 19.

Allard, P. 1955. Anomalies dans les temperatures de l'eau de mer observées au Cabo Frio au Brésil. Comité Central d'Oceanographie et d'Étude des Côtes, Service Hydrographique de la Marine. *Bulletin d'Information*, 2.

Allen, J. S. 1980. Models of wind-driven on the continental-shelf. *Annu. Rev. Fluid Mech.*, 12: 389-433.

Andrade, M. M. 2008. *Paleoprodutividade costeira da região de Cabo Frio, Rio de Janeiro, ao longo dos últimos 13.000 anos cal AP.* Thesis, Universidade Federal Fluminense, Niterói, Brazil, 275p.

André, D. L. 1990. *Análise dos parâmetros hidroquímicos na ressurgência de Cabo Frio.* Thesis, Universidade Federal Fluminense, Niterói, Niterói, Brazil, 203p.

Araujo, D. S. D. 1997. Cabo Frio Region. In: Davis, S. D., Heywood, V. H., Herrera-MacBryde, O., Villa-Lobos, J., Hamilton, A. C. (Eds.), *Centres of Plant Diversity: a guide and strategy for their conservation: The Americas.* WWF/IUCN, Oxford: 373-375.

Araujo, D. S. D. 2000. *Análise florística e fitogeográfica das restingas do Estado do Rio de Janeiro.* Thesis, Universidade Federal do Rio de Janeiro, 176p.

Araujo, D. S. D. and Maciel, N. C. 1998. Restingas fluminenses: biodiversidade e preservação. *Boletim FBCN*, 25: 27-51.

Barbiére, E. B. 1975. Ritmo climático e extração do sal em Cabo Frio. *Rev. Bras. Geografia*, 37: 23-109.

Barbiére, E. B. 1984. Cabo Frio e Iguaba Grande, dois microclimas distintos a um curto intervalo espacial. In: *Restingas: origem, estrutura, processos.* CEUFF, Niterói: 3-13.

Barbiére, E. B. 1986. Distribuição de pluviosidade ao longo do trecho Niterói-Cabo Frio (RJ). *Ciência e cultura*, 38: 766-767.

Barbiére, E. B. 1999. Origin and evolution of Quaternary coastal plain between Guaratiba and cape Frio, State of Rio de Janeiro, Brazil. In: Knoppers, B. A., Bidione, E. D. and Abrão, J. J. (Eds.). *Environmental Geochemistry of Coastal Lagoon System of Rio de Janeiro Brazil.* Série Geoquímica Ambiental, 6: 47-56.

Barbosa, D. S. 2003. *Sedimentação Orgânica na Lagoa Brejo do Espinho (Cabo Frio, RJ): Composição e Implicações Paleoclimáticas.* Thesis, Universidade Federal Fluminense, Niterói, 90p.

Braga, A. C. 2001. O ambiente e a biodiversidade, In: Costa, P. A. S. (Ed.) *Peixes marinhos do Estado do Rio de Janeiro.* FEMAR, SEMAD, Rio de Janeiro, 234p.

Calado, L., Silveira, I. C. A., Gangopadhyay, A., and Castro, B. M. 2010. Eddy-induced upwelling off Cape Sao Tome (22 degrees S, Brazil). *Cont. Shelf Res.*, 30: 1181-1188.

Campos, E. J. D., Gonçalves, J. E. and Ikeda, Y. 1995. Water mass characteristics and geostrophic circulation in the South Brazil Bight: summer of 1991. *Journal of Geophysical Research*, 18: 18537-18550.

Campos, E. J. D., Velhote, D. and Silveira, I. C. A. 2000. Shelf break upwelling driven by Brazil Current cyclonic meanders. *Geophys. Res. Let.*, 27: 751-754.

Castro, B. M. and Miranda, L. B. 1998. Physical oceanography of the Western Atlantic continental shelf located between 4° N and 34° S. In: Robinson, A. R., Brink, K. H. (Eds.), *The Sea*. John Wiley, New York: 209-251.

Coe, H. H. G. 2009. *Fitólitos como indicadores de mudanças na vegetação xeromórfica da região de Búzios / Cabo Frio, RJ, durante o Quaternário*. Thesis, Universidade Federal Fluminense, Niterói, 300p.

Coe, H. H. G. and Carvalho, C. N. 2013a. Cabo Frio - um enclave semiárido no litoral úmido do Estado do Rio de Janeiro: Respostas do clima atual e da vegetação pretérita. *GeoUSP*, 33: 136-151.

Coe, H. H. G., Alexandre, A., Carvalho, C. N., Santos, G. M., Silva, A. S., Sousa, L. O. F., and Lepsch, I. F. 2013b. Changes in Holocene tree cover density in Cabo Frio (Rio de Janeiro, Brazil): Evidence from soil phytolith assemblages. *Quaternary International*, 287:63-72.

Coelho-Souza, S. A., López, M. S., Guimarães, J. R. D., Coutinho, R., and Candella, R. N. 2012. Biophysical Interactions in the Cabo Frio upwelling system, Southeastern Brazil. *Brazilian Journal of Oceanography*, 60(3): 353-365.

Dourado, M. S. and Oliveira, A. P. 2001. Observational description of the Atmospheric and Oceanic Boundary Layers over the Atlantic Ocean. *Brazilian Journal of Oceanography*, 49 (1/2): 49-59.

Emilson, I. 1961. The Shelf and Coastal Waters of Southern Brazil. *Bol. Inst. Oceanogr.*, XI(2): 101-112.

Farág, P. R. C. 1999. *Estrutura do estrato arbóreo de mata litorânea semicaducifólia sobre solo arenoso no município de Búzios, RJ*. Thesis, Universidade Federal do Rio de Janeiro, Brazil, 87p.

Franchito, S. H., Rao, V. B., Stech, J. L., and Lorenzetti, J. A. 1998. The effect of coastal upwelling on the sea-breeze circulation at Cabo Frio, Brazil: a numerical experiment. *Ann. Geophys-Atmos. Hydro Space Sci.*, 16: 866-881.

Franchito, S. H., Rao, V. B., Oda, T. O., and Conforte, J. C. 2007. An observational study of the evolution of the atmospheric boundary-layer over Cabo Frio, Brazil. *Ann. Geophys.*, 25: 1735-1744.

Franchito, S. H., Oda, T. O., Rao, V. B., and Kayano, M. T. 2008. Interaction between coastal upwelling and local winds at Cabo Frio, Brazil: an observational study. *J. Appl. Meteorol. Climatol.*, 47: 1590-1598.

Ibraimo, M. M., Schaefer, C. E. G. R., Ker, J. C., Lani, J. L., Rolim-Neto, F. C., Albuquerque, M. A., and Miranda, V. J. 2004. Gênese e Micromorfologia de Solos sob Vegetação Xeromórfica (caatinga) na Região dos Lagos (RJ). *R. Bras. Ciência do Solo*, 28: 695-712.

INMET, 2005. *Normais climatológicas do Brasil, Período 1961-1990.*

Ireland, S. 1987. The Holocene sedimentary history of coastal lagoons of Rio de Janeiro state, Brazil. In: Tooley, M., Shennan, I. (Eds.), *Sea-level changes*. Basil Blackwell Ltd., Oxford: 25-66.

Kampel, M., Lorenzetti, J. A. and Silva Jr., C. L. 1997. Observação por satélite de ressurgências na costa S-SE brasileira. In: *Congresso Latino-americano sobre Ciências do Mar (COLACMAR)*, 7, Santos, SP, Brazil: 38-40.

Kousky, V. E., Kagano, M. T. and Cavalcanti, I. E. A. 1984. A review of Southern Oscillation: Oceanic-atmosferic circulation changes and related rainfal anomalies. *Tellus*, 36 A: 490-504.

Laslandes, B., Sylvestre, F., Sifeddine, A., Turcq, B., Albuquerque, A. L. S., and Abrão, J. 2006. Enregistrement de la variabilité hydroclimatique au cours des 6500 dernières années sur le littoral de Cabo Frio (Rio de Janeiro, Brésil). *C.R. Geosci.*, 338: 667-675.

Lima, H. C. 2000. *Leguminosas arbóreas da Mata Atlântica. Uma análise da riqueza, padrões de distribuição geográfica e similaridades florísticas do Estado do Rio de Janeiro.* Thesis, Universidade Federal do Rio de Janeiro, Brazil, 151p.

Lima, H. C. and Guedes-Bruni, R. R. 1997. Diversidade de plantas vasculares na Reserva Ecológica de Macaé de Cima. In: *Serra de Macaé de Cima: Diversidade e Conservação em Mata Atlântica* (H. C. Lima and R. R. Guedes-Bruni, eds.). Instituto de Pesquisas Jardim Botânico do Rio de Janeiro, Rio de Janeiro: 29-39.

Martin, L., Flexor, J. M. and Valentin, J. L. 1988. Influence du phénomène océanique pacifique, "El Niño" sur l'upwelling et le climat de la région de Cabo Frio, sur la côte brésilienne de l'État de Rio de Janeiro. *Comptes Rendus de l'Académie des Sciences de Paris*, série II, 307: 1101-1105.

Martin, L. and Suguio, K. 1989. Excursion route along brazillian coast between Santos and Campos. *International Symposyun on Global Changes in South America during the Quaternary*. Special Publication 2, São Paulo, Brazil, 122p.

Mascarenhas, A. S., Miranda, L. B. and Rock, N. J. 1971. A study of oceanographic conditions in the region of Cabo Frio. In: (Ed), C. J. (Ed.), *Fertility of the sea*. Gordon and Breach, New York: 285-308.

Miranda, L. B. 1985. Forma de correlação T-S de massas de água das regiões costeira e oceânica entre o Cabo de São Tomé (RJ) e a ilha de São Sebastião (SP), Brasil. *Boletim do Instituto Oceanográfico*, 33: 105-119.

Mooney, H. A., Bullock, S. H. and Medina, E. 1995. Seasonally dry tropical forests. Introduction, In: S. H. Bullock, Mooney, H. A., Medina, E. (Eds.), *Seasonally dry tropical forests*. Cambridge Univ. Press: 1-8.

Moreira da Silva, P. C. and Rodrigues, R. F. 1966. Modificações na Estrutura Vertical das Águas Sobre a Borda da Plataforma Continental por Influência do Vento. *Publicação do Inst. Pesq. da Marinha*, 13 p.

Moreira da Silva, P. C. and Rodrigues, R. F. 1973. A Ressurgência de Cabo Frio (I). *Publicação do Inst. Pesq. da Marinha*, 56 p.

Moreira da Silva, P. C. 1977. A ressurgência em Cabo Frio. *Publicação do Instituto de Pesquisas da Marinha* (IPqM),78.

Nimer, E. 1989. *Climatologia do Brasil*. 2 ed., IBGE, Rio de Janeiro.

Oda, T. O. 1997. *Influência da ressurgência costeira sobre a circulação local em Cabo Frio*. Thesis, INPE, São José dos Campos, Brazil.

Odum, E. P. 1983. *Ecologia*. Rio de Janeiro, 434 p.

Oliveira, A. C. C. 2008. *Registro de paleotemperaturas na plataforma continental de Cabo Frio, Rio de Janeiro, ao longo dos últimos 13.000 anos*. Thesis, Universidade Federal Fluminense, Niterói, Brazil, 205p.

Ortega, L. T. 1996. Variations paléohydrologiques et paléoclimatiques d'une région d'upwelling au cours de l'Holocene: enregistrement dans les lagunes cotières de Cabo Frio (État de RJ, Brésil), *Terre, Océan, Espace*, 6.

Palacios, J. R. 1993. *Estudo Espectral do fenômeno da Ressurgência de Cabo Frio (RJ, Brasil)*. Thesis, Observatório Nacional do CNpQ, Rio de Janeiro, Brazil.

Pelegri, J. L., Atistegui, J., Cana, L., Gonzales-Davila, M., Hernandez-Guerra, A., Hernandez-Leon, S., Marrero-Diaz, A., Montero, M. F., Sangra, P., and Santana-Casiano, M. 2005. Coupling between the open ocean and the coastal upwelling region off northwest Africa: water recirculation and offshore pumping of organic matter. *J. Mar. Syst.*, 54, p. 3-37.

Prado, D. E. 1991. *A critical evaluation of the floristic links between Chaco and Caatingas vegetation in South America*. Thesis, University Saint Andrews, 173 p.

Prado, D. E. 2000. Seasonally Dry Forests of Tropical South America: from forgotten ecosystems to a new phytogeographic unit. *J. Bot.* 37: 437-461.

Rizzini, C. T. 1979. *Tratado de Fitogeografia do Brasil.* EDUSP, São Paulo.

Rodrigues, R. R. and Lorenzetti, J. A. 2001. A numerical study of the effects of bottom topography and coastline geometry on the Southeast Brazilian coastal upwelling. *Cont. Shelf Res.*, 21: 371-394.

Signorini, S. R. 1978. On the Circulation and the Volume Transport of the Brazil Current Between the Cape of São Tomé and Guanabara Bay. *Deep Sea Research*, 25: 453-443.

Scheel-Ybert, R. 2000. Vegetation stability in the Southeastern Brazilian coastal area from 5500 to 1400 ^{14}C yr BP deduced from charcoal analysis. *Review of Palaeobotany and Palynology*, 110: 111-138.

Stech, J. L., Lorenzzetti, J. A., 1992. The response of the South Brazil Bight to the passage of wintertime cold fronts. *Journal of Geophysical Research*, 97: 9507-9520.

Stramma, L. and Peterson, R. G. 1990. The South-Atlantic current. *J. Phys. Oceanogr.*, 20: 846-859.

Stramma, L. and England, M. 1999. On the water masses and mean circulation of the South Atlantic Ocean. *J. Geophys. Res.*, 104: 20863-20883.

Torres Jr., A. R. 1995. *Resposta da ressurgência costeira de Cabo Frio a forçantes locais.* Thesis, COPPE Universidade Federal do Rio de Janeiro, Brazil.

Turcq, B., Knoppers, B., Bidone, E. D., and Abrão, J. J. 1999. Origin and evolution of the Quaternary coastal plain between Guaratiba and Cabo Frio, State of Rio de Janeiro, Brazil In: Knoppers, B., Bidone, E. D., Abrão, J. J. (Eds.), *Environmental Geochemistry of Coastal Lagoon Systems*, Rio de Janeiro, Brazil. CEUFF, Niterói: 25-46.

Ule, E. 1967. A vegetação de Cabo Frio. *Boletim Geográfico*, 200: 21-32.

Ururahy, J. C. C. 1987. Nota sobre uma formação fisionômico-ecológica disjunta da estepe nordestina na área do Pontal de Cabo Frio, RJ. *Rev. Bras. Geogr.*, 49: 25-29.

Valentin, J. L. 1984. Analyse des paramètres hydrobiologiques dans la remontée de Cabo Frio (Brésil). *Marine Biology*, 82: 259-276.

Valentin, J. L., André, D. L. and Jacob, S. A. 1987. Hidrobiology in the Cabo Frio (Brazil) upwelling: two-dimensional structure and variability during a Wind cycle. *Continental Shelf Research*, 7: 77-88.

Valentin, J. L. and Coutinho, R. 1990. Modelling maximum chlorophyll in the Cabo Frio (Brazil) upwelling: a preliminary approach. *Ecological Modelling*, 52: 103-113.

Valentin, J. L. 1994. A ressurgência. Fonte de vida nos oceanos. *Revista Ciência Hoje*, 18: 19-25.
Winant, C. D. 1980. Coastal circulation and wind-induced currents. *Annu. Rev. Fluid Mech.*, 12: 271-301.

In: Upwelling   ISBN: 978-1-62948-174-6
Editors: W. E. Fischer and A. B. Green   © 2013 Nova Science Publishers, Inc.

Chapter 2

NAMIBIAN UPWELLING AND ITS EFFECTS ON MACROZOOBENTHIC DIVERSITY

Michael L. Zettler and Falk Pollehne*
Leibniz Institute for Baltic Sea Research Warnemünde (IOW),
Rostock, Germany

ABSTRACT

The Namibian sector of the southern Atlantic shelf is part of the Benguela upwelling system and among the most productive marine ecosystems on our planet. The primary production is significantly enhanced by upwelling of nutrient-rich deep water further charged by nutrients recycled from shelf sediments. Subsurface shelf waters of the Benguela system are characteristically low in oxygen content and periodically anoxic. Free hydrogen sulphide in bottom shelf water has frequently been observed during austral spring and summer and occasional upwelling of sulfidic water at the coast has disastrous impacts on higher life forms. The sedimentary organic carbon load is high compared to oxygen availability resulting in deficiency of dissolved oxygen in bottom waters. These harsh environmental conditions normally hamper the development of higher benthic life. Macrozoobenthos were investigated in this area during the years 2004, 2008 and 2011 at 52 stations between 25 and 2513 m water depths. Altogether 134 grab samples were analyzed. Surprisingly, at all stations and all times macrozoobenthos abundance could be observed, even in oxygen minimum zones (OMZ) (bottom waters with oxygen contents below 0.5

---

* E-mail: michael.zettler@io-warnemuende.de.

and 0.1 ml/l respectively). Although the biodiversity decreased in the center of the upwelling cells (at water depths of around 50 to 400 m), a high number of well-adapted species remained to form typical benthic communities. Particularly at the fringes of the OMZ biomass an abundance of these species–reduced communities can maintain high values that are comparable to those in oxic environments.

As a specific feature of the Benguela System the upwelling cells constitute barriers for species diversity but not for biomass or abundance. This may be an effect of the stability of this system over millions of years, which could have promoted the adaptation of macrozoobenthic communities to extreme environmental conditions.

INTRODUCTION

Upwelling of nutrient-rich water in regions with sufficient solar irradiation stimulates the growth and reproduction of primary producers. Upwelling zones can be identified by cool sea surface temperatures and high concentrations of chlorophyll a (e.g. Sarhan et al. 1999) and are preferentially situated at western coasts of the continents. Worldwide, there are five major coastal currents associated with upwelling areas: the Canary Current (off Northwest Africa), the Benguela Current (off Southwest Africa), the California Current (off California and Oregon), the Humboldt Current (off Peru and Chile), and the Somali Current (off Somalia and Oman). In these regions oxygen minimum zones (OMZ) (<0.5 ml l^{-1} dissolved oxygen) develop on the continental slope. The oxygen concentrations in the upwelling areas vary somewhat seasonally and between coastal areas. In most of the upwelling regions the OMZ is a phenomenon of deeper water layers (over several 100 meters) in waters off Namibia (Benguela Current) the OMZ ranges between 50 and 400 m. The Benguela upwelling system ranges from Angola to South Africa in the southeast Atlantic and is bordered in the north by the tropical Angola Current, where warm and cold water masses converge at about 16° S at the Angola–Benguela Frontal Zone (ABFZ) (Veitch et al. 2006, Bodungen et al. 2008, Mohrholz et al. 2008). Although the northern Benguela system off Namibia (focus of the present study) has much in common with its equivalent in the South Pacific, the southern part is rather different because it is bordered to the south by the warm retroflection zone of the Agulhas Current (Shannon 1989). The ancient Benguela system evolved in the mid to late Miocene epoch (about 12 million years ago) and the Benguela system as we know it today is about 2 million years old. The Benguela upwelling system exists since that time, with interruptions during the Pleistocene, and the biota could adapt to this unique environment. In general OMZs generate benthic ecosystems that differ fundamentally from those in well-oxygenated environments (Levin 2003). The upwelling system off Namibia may be regarded as an intermediate system, combining high productivity typical for upwelling areas with the sporadic presence of sulphide in the water column (Borchers et al. 2005). The presence of low-oxygen water and sulphidic anoxic sediments in coastal waters of the central Namibian Shelf is well documented (e.g. O'Toole 1997) and is primarily believed to regulate benthic diversity patterns. Since Sanders (1969), who first explored benthic communities in OMZ regions off Walvis Bay, West Africa, several studies have been carried out in the eastern Pacific Ocean (e.g. Mullins et al. 1985, Gutiérrez et al. 2000, Levin et al. 2002, Gallardo et al. 2004, Palma et al. 2005, Quiroga et al. 2005) and the Arabian Sea (e.g. Smith et al. 2000, Cowie & Levin 2009, Gooday et al. 2009, Hughes et al. 2009, Levin et al. 2009, Ingole et al. 2010). A recent study of the Benguela system (Scott et al. 2012) summarizes the knowledge on macrozoobenthic diversity in water depths of <15 m. However, Benthic communities in offshore subtidal regions have been studied only rarely (Sakko 1998). For the northern part of the Benguela upwelling system some papers on macrozoobenthos distribution and its relation to the OMZ and other environmental parameters were recently published by the authors and form the basis of this chapter (Bochert & Zettler 2013, Zettler et al. 2009, Zettler et al. 2013). The aim of this study is to describe the benthic diversity and abundance patterns in its relation to oxygen deficiency along the Namibian coast.

MATERIAL AND METHODS

Study Area

Namibia's coastline extends to about 1500 km in north-south direction on the south-east African coast (Bianchi et al. 1999). Hydrological and pelagic biological patterns (i.e. pelagic fish larvae, zooplankton biomass) of the Angola-Benguela frontal zone were described recently in detail by Postel et al. (2007), Mohrholz et al. (2008), Bodungen et al. (2008) and Ekau et al. (2010). Due to the high productivity the sediments are predominantly of biogenic origin (Borchers et al. 2005). Results presented in this study combine data from three cruises conducted in the vicinity of the ABFZ between 17 and 23°S along the Namibian coasts. Benthic samples and environmental data collected at 52 locations off the Namib zoogeographical province (sensu Emanuel et al. 1992) were analyzed to describe biodiversity patterns of macrozoobenthic species (Figure 1).

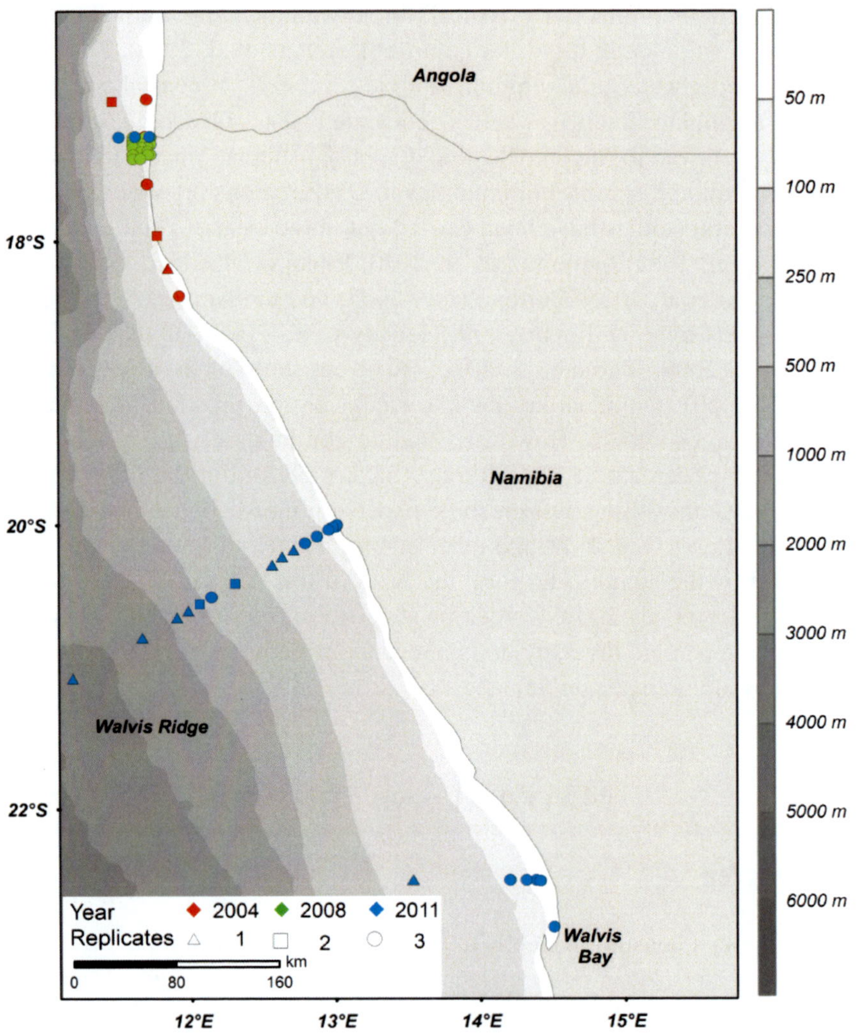

Figure 1. Investigation sites within the Benguela upwelling area in northern Namibia between 2004 and 2011.

Sampling

In the years 2004, 2008 and 2011 and between 17° S and 22.8° S, a total of 52 stations with 134 samples were analyzed with comparable effort and methodology (see Figure 1 for locations) for both abiotic and biotic variables.

Sampling was performed on three cruises in 2004 (RV Alexander von Humboldt), 2008 and 2011 (both RV Maria S. Merian). Depending on the water depth a 0.1 m² van Veen grab (25 to 204 m) or a 0.1 m² Reineck box corer (306 to 2513 m) was used repeatedly (Table 1, Figure 2). One, two or three replicate benthic samples were taken at each station and handled separately (Table 1). Altogether 134 benthic samples were collected. All samples were sieved through a 1 mm² screen and animals were preserved on board in 4% buffered formaldehyde. Sorting was conducted at the laboratory using a stereomicroscope with 10–200x magnification. All macrofauna samples were identified to the lowest possible taxonomic level. The nomenclature was checked using the World Register of Marine Species (WoRMS: http://www.marinespecies.org/index.php).

Figure 2. Equipment used for macrozoobenthos sampling during cruises in 2004, 2008 and 2011 in waters off Namibia. Left: 0.1 m² Reineck box corer used for water depth between 306 and 2513 m; right: 0.1 m² van Veen grab used for shallow water sampling down to 204 m depth.

                    Michael L. Zettler and Falk Pollehne

Table 1. List of stations of the present study. The number of replicates and the employed sampling gear is indicated (vV=0.1 m² van Veen grab; BC=0.1 m² box corer). Characteristic environmental variables in bottom water and sediment (Oxy=oxygen, Org=organic content) are given

station	date	depth in m	longitude E	latitude S	replic.	gear	Org % wt	Oxy ml l⁻¹
BE1-2004	12.05.2004	42	11.922	18.385	3	vV	-	-
BE2-2004	12.05.2004	32	11.841	18.191	1	vV	-	1.51
BE3-2004	12.05.2004	45	11.768	17.961	2	vV	-	-
BE4-2004	13.05.2004	60	11.700	17.600	3	vV	-	0.08
BE5-2004	13.05.2004	65	11.683	17.263	3	vV	-	0.05
BE6-2004	13.05.2004	29	11.695	17.002	3	vV	-	0.78
BE8-2004	13.05.2004	117	11.458	17.017	2	vV	-	0.51
BE011-2008	04.03.2008	82	11.658	17.289	3	vV	12.67	0.32
BE01-2008	04.03.2008	106	11.603	17.268	3	vV	7.13	0.31
BE012-2008	04.03.2008	102	11.603	17.290	3	vV	7.53	0.39
BE03-2008	04.03.2008	64	11.687	17.268	3	vV	10.21	0.51
BE06-2008	04.03.2008	31	11.724	17.267	3	vV	4.22	0.83
BE09-2008	04.03.2008	49	11.710	17.289	3	vV	8.99	0.53
BE14-2008	04.03.2008	114	11.601	17.316	3	vV	8.53	0.12
BE16-2008	05.03.2008	83	11.659	17.316	3	vV	7.57	0.24
BE17-2008	05.03.2008	26	11.723	17.316	3	vV	-	1.21
BE18-2008	05.03.2008	56	11.704	17.339	3	vV	9.12	0.88
BE20-2008	05.03.2008	84	11.659	17.339	3	vV	8.55	0.33
BE23-2008	05.03.2008	114	11.602	17.340	3	vV	5.53	0.06
BE26-2008	05.03.2008	82	11.659	17.368	3	vV	11.49	0.4
BE29-2008	05.03.2008	53	11.704	17.368	3	vV	9.24	0.67
BE30-2008	05.03.2008	30	11.724	17.390	3	vV	-	1.19
BE33-2008	06.03.2008	58	11.704	17.390	3	vV	7.50	0.6
BE39-2008	06.03.2008	117	11.602	17.390	3	vV	7.76	0.06
BE41-2008	06.03.2008	117	11.602	17.421	3	vV	6.40	0.07
BE42-2008	06.03.2008	93	11.652	17.420	3	vV	12.48	0.14
KU3-2011	15.08.2011	150	11.501	17.267	3	vV	5.25	0.83
KU4-2011	15.08.2011	102	11.615	17.264	3	vV	8.83	0.66

station	date	depth in m	longitude E	latitude S	replic.	gear	Org % wt	Oxy ml l^{-1}
KU5-2011	15.08.2011	39	11.717	17.262	3	vV	9.88	0.82
NAM004-2011	24.08.2011	133	12.710	20.175	1	vV	4.12	0.48
NAM005-2011	24.08.2011	154	12.629	20.223	1	vV	11.64	0.42
NAM006-2011	24.08.2011	204	12.560	20.281	1	vV	12.05	0.55
WFB1-2011	26.08.2011	27	14.503	22.830	3	vV	12.76	0.06
NAM-BE03-2011	27.08.2011	33	13.004	20.010	3	vV	10.08	0.49
NAM-BE04-2011	27.08.2011	30	13.010	19.998	3	vV	1.02	0.51
NAM-BE01-2011	30.08.2011	56	12.969	20.022	3	vV	7.71	0.59
NAM-BE02-2011	30.08.2011	44	12.984	20.013	3	vV	5.88	0.68
NAM009-2011	01.09.2011	306	12.304	20.413	2	BC	14.27	0.48
NAM011-2011	01.09.2011	403	12.141	20.509	3	BC	8.08	0.62
NAM002-2011	06.09.2011	101	12.872	20.079	3	vV	26.09	0.55
NAM003-2011	06.09.2011	123	12.791	20.127	3	vV	28.80	0.49
NAM012-2011	07.09.2011	550	12.060	20.556	2	BC	12.93	1.03
NAM013-2011	07.09.2011	707	11.979	20.604	1	BC	15.80	2.08
NAM014-2011	07.09.2011	866	11.898	20.652	1	BC	15.80	2.81
NAM001-2011	08.09.2011	66	12.954	20.032	3	vV	25.41	0.55
NAM018-2011	10.09.2011	2513	11.164	21.081	1	BC	7.82	5.28
NAM016-2011	13.09.2011	1380	11.653	20.795	1	BC	14.34	4.33
NAM-BE17-2011	18.09.2011	35	14.381	22.500	3	vV	0.47	0.51
NAM-BE18-2011	18.09.2011	54	14.314	22.500	3	vV	12.23	0.25
NAM-BE19-2011	18.09.2011	82	14.201	22.500	3	vV	24.37	0.15
NAM-BE20-2011	18.09.2011	25	14.409	22.506	3	vV	0.54	0.22
NAM-BE21-2011	18.09.2011	156	13.533	22.499	1	vV	13.10	-

Environmental variables such as salinity, temperature and oxygen concentrations in the water column were recorded, with special attention to the benthic boundary layer, by means of a profiling CTD-system (Seabird, USA) with attached oxygen sensor (Seabird, USA) and a 13 bottle sampling rosette. Oxygen sensors were calibrated on board by immediate potentiometric Winkler titration of three samples per water column, including the closest position to the sediment. No titrated oxygen measurement exists for the three samples.

An additional sediment sample was taken by grab to extract the upper surface sediment layer ($\leq$20 mm) for analyses of median grain size (laser particle sizer Cilas 1180L) and organic matter estimation by weight loss upon ignition at 500°C.

Table 1 displays values for bottom water and sediment variables of stations sampled during this study. At some stations with extensive layers of coarse shell debris sediment analyses were not possible.

Statistical Analyses

Independent of taxonomical resolution all taxa were selected for analysis. The identification was targeted to the species or genus level. In cases where this was impossible, identification was adjusted to the family level. For some groups (e.g. Bryozoa, Hydrozoa, Nemertea, Anthozoa, and Turbellaria) no taxonomical expertise was available. In these cases and for damaged material and juveniles higher taxonomical levels (order, class etc.) were employed. The allocation to taxonomic levels is shown in Tables 2 and 3.

For each benthic sample the abundance and biomass (wet weight) of each taxon was counted and weighed, respectively. The abundance and biomass per square meter of all individuals of each sample (no pooling) was calculated.

Table 2. Taxonomical solution of the taxa of 134 analyzed grab samples off Namibia

	number	cumulative %
species level	152	29
genus level	361	69
family level	441	84
taxa	525	100

Table 3. Taxonomical solution of the data used within the analysis of macrozoobenthic diversity off Namibia

	number	%
data	2614	100
species level	1122	43
genus level	1829	70
family level	2322	89
class level	2614	100

RESULTS

Whereas the very shallow stations (20 to 30 m water depth) were relatively well oxygenated with values above 1 ml O_2/l, the near bottom water oxygen content of the mid-water stations (30 to 400 m) decreased rapidly to below 1 and 0.5 ml/l (Figure 3a). At few stations even lower concentrations down to 0.05 ml O_2/l were measured. Below 400 m water depth the oxygen content of the bottom layer increased slowly from around 0.5 ml/l to more than 5 ml/l in the deep sea (2513 m water depth) (Figure 3a). The near bottom water temperature decreased from 18°C in shallow depths to only 2.8°C in the deepest part of the transect (Figure 3b). The number of species in the grab samples is directly linked to the oxygen content of the bottom water, high at the shallow stations, low to very low in the middle range and increasing again towards the deep water stations (Figure 3c, Figure 3d). Although the macrobenthic diversity is depleted specifically in the depth range between 30 and 200 m and generally at oxygen contents below 0.5 ml/l, macrofauna abundance and biomass reveal different patterns. Even at stations with very low oxygen content (<0.1 ml/l) both the abundance and the biomass can reach values comparable to those in oxic environments. Especially for the biomass this is very obvious (Figure 3e-h).

In 134 samples from the study area 525 species/taxa of 27 animal classes could be identified (Figure 4). The most diverse and abundant classes were the Polychaeta, Malacostraca, Gastropoda and Bivalvia followed by some echinoderm classes. Another very abundant class were the Oligochaeta; however, due to the lack of expertise in taxonomy, they could not be identified to the species level. Additionally, due to sieving through 1.0 mm, the completeness of this group (of small taxa) may be questionable.

Michael L. Zettler and Falk Pollehne

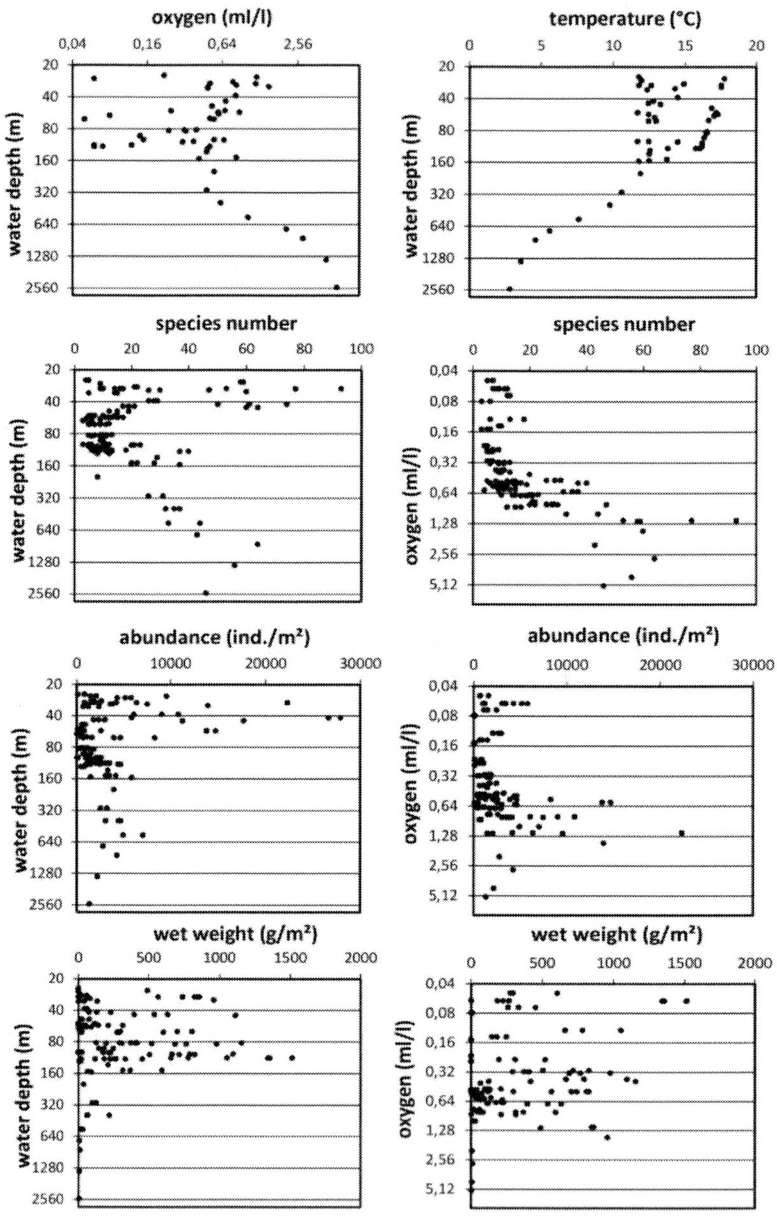

Figure 3. Relations of water depth, near bottom water oxygen, temperature and some macrozoobenthic parameters to each other from 134 samples in the Benguela upwelling area of northern Namibia sampled between 2004 and 2011.

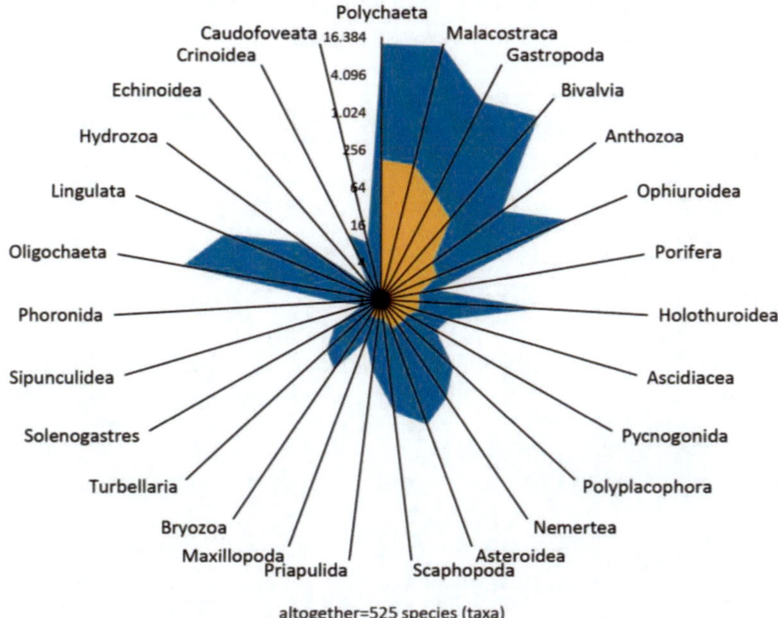

Figure 4. Species number (orange) of taxonomical classes and their cumulative abundance (blue) per square meter gathered from 134 samples in the Benguela upwelling area between 2004 and 2011. Please note the logarithmic scale of the y-axis.

As expected, the faunal composition varied extremely depending on water depth, oxygen content and sediment structure (Figure 5 and 6), which are again all dependent on the water mass distribution and dynamics. Whereas in the shallowest and most exposed stations (water depths between 20 and 30 m) shell gravel (e.g. from the brachiopod *Discinisca tenuis* (Sowerby)) or well sorted sands exist (Figure 5a), the sediment was muddier and richer in organic content in the deeper parts. The relative well oxygenated ($O_2>1$ ml/l) *Discinisca* shells were colonized by a very biodiverse benthic fauna dominated by *D. tenuis* itself. Other examples of very abundant species were the polychaete *Dipolydora* cf. *normalis* (Day, 1957), the ophiurids *Ophiolepis paucispina* (Say, 1825) and *Ophiactis luetkeni* Marktanner-Turneretscher, 1887, the tanaidacean *Calozodion dominiki* Bochert, 2012 and the amphipod *Melita zeylanica* Stebbing, 1904. Already a few meters below 30 m of water depth the sediment composition changed abruptly from sandy to muddy sand and poor mud. In the mud belt from 40 to 400 m depth the organic content can reach values of up to 20% of the dry weight. The mud content (fraction <63 µm) is above 80 or 90%. The benthic faunal composition varies in dependence

　Michael L. Zettler and Falk Pollehne

of the region of the upwelling area. At the northern margin a bivalve-gastropod dominated community occurs (Figure 5 d).

Figure 5. Images of the residuals (after sieving with 1 mm) of grab and box core samples of the Benguela upwelling area in northern Namibia from the depth range between 30 and 866 m. (A=BE30-2008, B=NAM-BE03-2011, C=NAM-BE01-2011, D=BE011-2008, E=NAM003-2011, F=NAM005-2011, G=NAM012-2011, H=NAM014-2011).

Figure 6. Most frequent and dominant taxa of the Benguela upwelling area in northern Namibia from the depth range between 50 and 150 m. a) *Diopatra neapolitana capensis* Day, 1960; b) *Tectonatica sagraiana* (d'Orbigny, 1842); c) *Aslia* cf. *spyridophora* (H.L. Clark, 1923); d) *Nassarius vinctus* (Marrat, 1877); e) *Iphinoe africana* Zimmer, 1908; f) *Sabellaria eupomatoides* Augener, 1918; g) *Paraprionospio pinnata* (Ehlers, 1901); h) *Sinupharus galatheae* Cosel, 1993; i) *Pseudonereis variegata* (Grube, 1857); j) *Nuculana bicuspidata* (Gould, 1845).

Long-living species like the bivalve *Nuculana bicuspidata* (Gould, 1845) (Figure 6j) and the gastropod *Nassarius vinctus* (Marrat, 1877) (Figure 6d) dominate this area. Other frequent co-occurring species are the gastropods *Tectonatica sagraiana* (d'Orbigny, 1842) (Figure 6b) and *Bullia skoogi* (Odhner, 1923), the polychaetes *Sabellaria eupomatoides* Augener, 1918 (Figure 6f), *Pseudonereis variegata* (Grube, 1857) (Figure 6i) and *Paraprionospio pinnata* (Ehlers, 1901) (Figure 6g). In the center of the upwelling area more mobile and short-living species communities exist (Figure 5b, c). Cumaceans like *Iphinoe africana* Zimmer, 1908 (Figure 6e) and *Upselaspis caparti* (Fage, 1951), the holothurian *Aslia* cf. *spyridophora* (H.L. Clark, 1923) (Figure 6c) and the bivalve *Sinupharus galatheae* Cosel, 1993 (Figure 6h) are characteristic for this region. Polychaetes like *Sigambra* cf. *robusta* (Ehlers, 1908), *Pseudonereis variegata* (Grube, 1857) (Figure 6i) and *Paraprionospio pinnata* (Ehlers, 1901) (Figure 6g) are a frequent co-occurring species as well. Interjacent a depauperate community is dominated by the same cumaceans but with large quantities of shell gravel of the bivalve *Lucinoma capensis* (Thiele & Jaeckel, 1931) (Figure 5e).

Within this mud belt in the depth of around 150m a shell gravel zone covered with mud was detected. The sieved residuals are mainly dominated by subrecent shells of a formerly species-rich community and not comparable to recent colonization patterns (Figure 5f). Below this layer biodiversity increases with water depth. The deep water communities between 400 and 1000 m are dominated by several polychaete species of the genus *Prionospio, Paraonides, Levinsenia, Spiophanes, Cirrophorus* and *Diopatra* and sea pens (e.g. *Virgularia* sp.) (Figure 5g, h).

Other families of polychaetes are for example represented by the Eunicidae, Lumbrineriidae, Onuphidae, Orbiniidae and Maldanidae. Additionally some mollusc species of the genus *Cadulus, Dosinia, Yoldiella* and *Limatula* occur frequently. Whereas the sediment characteristics between 300 and 800 m could be described as sandy mud, the composition changes in deeper waters. At all stations from 866 m to 2513 m the sediment is mainly composed of radiolarian ooze (Figure 7).

The two deepest stations (1380 and 2513 m respectively) are also dominated by polychaetes like *Ophelia* sp., *Ophryotrocha* sp., *Prionospio* sp., *Spiophanes* sp. and *Paraonis* sp. Other co-occurring species with lower abundances belong to *Dondersia* sp. (Solenogastres), *Chaetoderma* sp. (Caudofoveata), *Leptochelia* sp. and *Paratanais* sp. (both Tanaidacea).

Figure 7. The sediment of the deep water stations between 866 and 2513 m consists mainly of radiolarian ooze. The image comes from NAM014-2011. Beside the radiolarian fragments a pectinid bivalve, polychaete tubes and foraminifera are visible.

DISCUSSION

Besides the upwelling of the Humboldt and the Somali current systems the Benguela current belongs to the largest and most persisting upwelling systems. Although the appearance of oxygen minimum zones is not restricted to upwelling areas, they are generally characteristic features of these systems (Levin et al. 2009, Sakko 1998). Benthic biodiversity is affected by the continuous or periodical harsh environmental conditions that go along with the surplus provision of organic matter to the seafloor. Whereas under normal oceanic conditions higher heterotrophic organisms seek to accumulate organic matter, in these regions they have to adapt to the adverse effects of cornucopia. Periods of anoxic conditions and sulphide poisoning, very often accompanied by toxic algae blooms, lead to mass mortalities of organisms, which can often be observed on the coastline of the Benguela region (Levin et al. 2009). Here, the line between extremely favorable growth conditions and poisonous environmental effects is particularly fine and therefore the pressure for adaptation is accordingly high. Large taxa are more sensitive than small taxa to hypoxia (Levin et al. 2009) and so a change in the size spectrum can be expected. Large mats of filamentous sulphide oxidizing bacteria (see Figure 8) probably provide detoxified microhabitats for eukaryotic benthic species under suboxic conditions (Bernhard et al. 2000), which then get rid of poisonous sulphide but still need physiological adaptations to low or missing oxygen values.

Figure 8. The sediment within the OMZ range is characterized by green mud and large mats of filamentous sulphide oxidizing bacteria (e.g. *Beggiatoa* sp.).

Some protozoans have solved this problem by switching to alternative oxidants like NO_3^- (Risgaard-Petersen et al. 2006, Högslund et al 2008, Leiter & Altenbach 2010), which was up to recently just thought to be reserved to microbial physiology. Which species are able to adapt to these conditions in the different upwelling areas additionally depends on the regions. If the oxygen content is still tolerable for metazoans, especially small, soft-body invertebrates (e.g. oligochaetes and polychaetes) with opportunistic life history traits and small generation times predominate. Some perennial mollusc species and families seem to be able to adapt to low and suboxic conditions as well (see review of Levin et al. 2009).

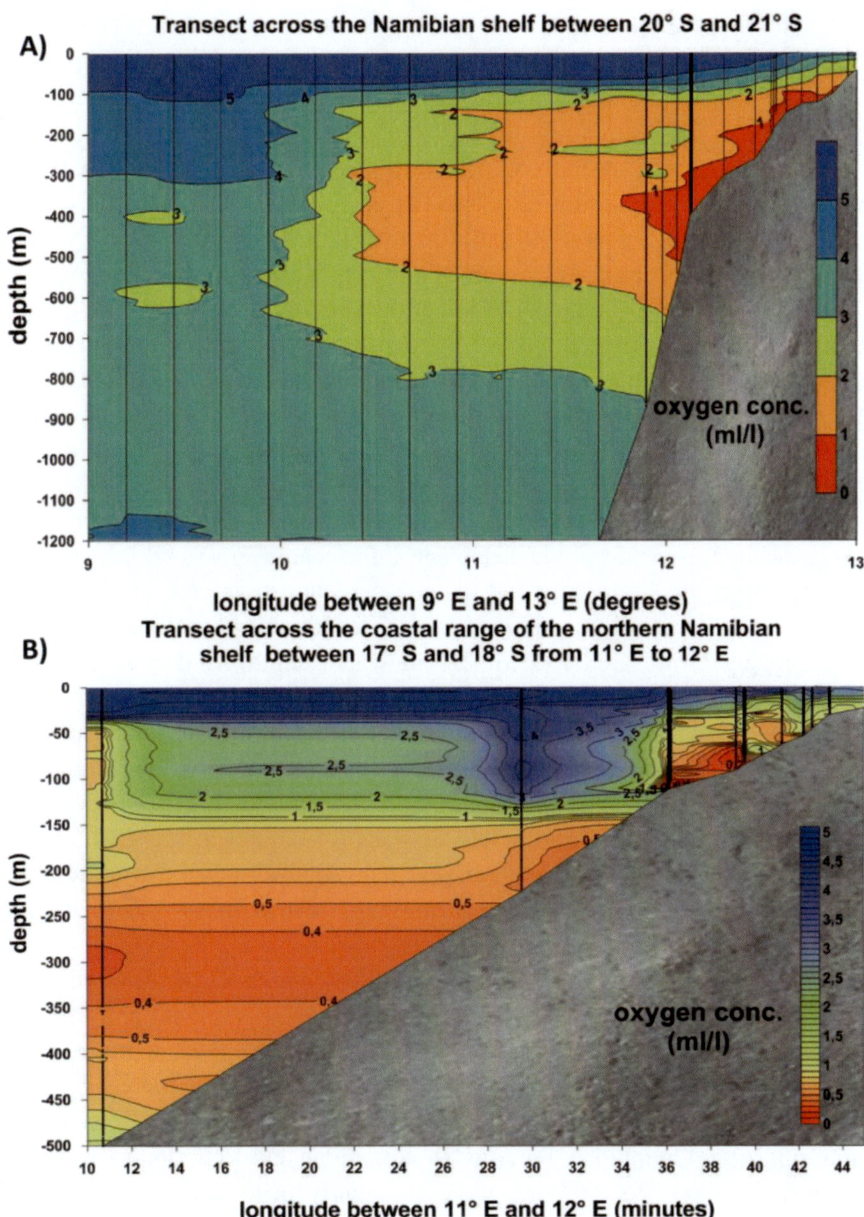

Figure 9. A) Isolines of oxygen concentration off the Namibian coast on a series of transects from 9° E to 13° E between 20° and 21° S in September 2011. Vertical lines indicate CTD-stations. B) The extremely low oxygen concentrations in the bottom water at 100 m depth in the northern part of the Benguela at 17° S in March 2008.

Their specific occurrence is probably not only related to their basic physiological capabilities but also to the age of the system and the time, that was given to endemic organisms, to adapt to the specific dynamics of anoxia and hypoxia in the distinct regions.

In the Benguela system of northern Namibia the hypoxic and anoxic impact on benthic communities due to the upwelling is mainly focused on the water depth range between 50 and 400 m (Figure 9a); often with extremely low oxygen concentrations in the shallower range (Figure 9b). Although sulphidic blowing-up events could affect the shallower areas as well (Emeis et al. 2004), the more exposed and well-structured sea bottom near-shore (<30 m water depth) shows higher benthic diversity. In the core of the OMZ (50 to 400 m water depth) the diversity decreases rapidly. However the abundance and biomass can reach similar values as in oxic waters. Two to three different macrobenthic communities can be discerned at the fringe of the OMZ, where they are dominated by bivalves and gastropods, and in the core area by short-living and opportunistic polychaetes and cumaceans. However, depending on the length and the strength of the oxygen depletion bivalves, anthozoans and holothurians can also occur in high abundance. In general the benthic fauna of the Benguela OMZ area (50 to 400 m water depth) is dominated by the families of Bodotriidae, Cucumariidae, Nassariidae, Nuculanidae, Paraonidae, Pilargidae, Sabellariidae and Spionidae. These findings are in accordance with observations of other upwelling areas (e.g. Gallardo et al. 2004, Palma et al. 2005, Cowie & Levin 2009, Gooday et al. 2009, Levin 2003, Levin et al. 2009a, b). Higher taxonomic levels of species adapted to the OMZ regions seem to be the same, such as families (e.g. Aphroditidae, Cirratulidae, and Nereidae) or in some cases genera (e.g. *Cossura* sp., *Diopatra* sp., *Paraprionospio* sp., *Pectinaria* sp.).

The lower fringe of the OMZ area of the Benguela current system can be fixed in the range of 300 to 400 m water depth with seasonal and long term variations. Continuously increasing oxygen values support a more diverse benthic colonization and the macrobenthic biodiversity is increasing again. The depth zone between 400 and 600 m is dominated by the families Lumbrineridae, Onuphidae, Paraonidae, Spionidae and Veneridae. The deepest stations, from 600 to 2500 m, were frequently occupied by the following families: Paraonidae, Pectinidae, Spionidae and Tanaissuidae.

Although the similarities mentioned above between macrozoobenthic colonization of the Benguela OMZ and other upwelling areas exist, the waters off northern Namibia accommodate numerous faunal components, which characterize this ecosystem as unique. Especially the mass occurrence of long living bivalves *Nuculana bicuspidata* and *Sinupharus galatheae* and the gastropod *Nassarius vinctus* in those parts of the OMZ with extremely low oxygen concentrations below 0.3 ml O_2 l^{-1} are conspicuous and may be based on very specific physiological pathways or symbiotic interactions. Whereas this *Nuculana-Nassarius* community probably experiences episodic anoxia and extremely low ambient oxygen concentrations (see Zettler et al. 2009), the cumacean community (dominated by *Iphinoe africana* and *Upselaspis caparti*) is situated in the center of the mud belt with long lasting periods of anoxia and high sulphide concentrations in the water column (Emeis et al. 2006). These two cumacean species were already documented for the Benguela current system in the results of the Danish "Galathea"-Expedition in the 1950s (Jones 1955, 1956), and later by Bochert & Zettler (2011). Oxygen levels generally become hypoxic to anoxic during the late austral summer to autumn and ventilation occurs during the late winter to spring when fresher Eastern South Atlantic Central Water (ESACW) is drawn onto the shelf (Mohrholz et al. 2008). In general, even during this winter replenishment phase, the oxygen values in the bottom water are very low due to the high sedimentary oxygen demand. We think that this seasonal change of oxygen/sulphide conditions enables a reduced 'pioneer' community of opportunistic species like these two cumaceans to colonize these areas quickly from adjacent regions (Zettler et al. 2013).

At some stations in the core of the OMZ between 100 and 150 m depth rich and diverse molluscan shell deposits exist (Figure 5) in layers a few centimeters to decimeters deep in the sediment. These well preserved deposits document periods of reduced organic input or improved oxygen supply for at least several years or decades within the last centuries.

Macrofauna diversity patterns are reflecting the recent hydrographical and biogeochemical conditions in this area, such as the extent and strength of upwelling processes, the boundaries of OMZs and the limits of major water masses. The combination of the assessment of recent and historical macrofauna distribution patterns may therefore form a basis for the reconstruction of historic conditions over past climate cycles (Michel et al. 2011). This way, research on macrofauna diversity in hitherto unmapped coastal systems can feed directly into regional climate reconstruction, which forms a basis for future climate prediction.

The environmental and faunal differences between the OMZ core and the upper and lower fringes are obvious. Oxygen availability appears to be the driver for diversity in all groups of benthic fauna (Diaz & Rosenberg 1995). Nevertheless, other environmental factors (e.g. sediment characteristics and structure) are additional parameters influencing benthic communities in OMZ areas and at the margins as well (e.g. Gooday et al. 2010). Biogenic structures (e.g. shell gravel) and activities (e.g. sulphide oxidizing filamentous bacteria) provide, on small spatial scales, substantial habitat heterogeneity within the OMZs. Although the alpha-diversity (species richness) is impacted significantly by hypoxia and anoxia phases, the density and biomass can reach values as high as those in comparable oxic waters. Being in operation for 2 million years, the Benguela current upwelling system may have provided more time, than other comparable upwelling areas, for evolutionary adaptation to its harsh environmental conditions and may therefore contain extremely well adapted species, communities and symbiotic interactions. In the evolution of benthic organisms upwelling areas with its OMZ regions generally have a very high importance as (1) barriers for distribution pathways of benthic and pelagic reproductive stages of sedentary life forms, (2) drivers for evolutionary development by creating strong gradients in selective pressures and barriers to gene flow for some taxa (e.g. Foraminifera, sensu Gooday et al. 2010), (3) catalyzer in the selection of well adapted species to hypoxia and (4) origin of symbiotic arrangements between metazoans and prokaryotes (Bernhard et al. 2000). This constitutes the important role of upwelling OMZs in the world ocean as hotspots of element turnover, biological diversity and evolutionary potential.

ACKNOWLEDGMENTS

We are grateful to Deutsche Forschungsgemeinschaft (DFG) for the provision of ship time and to the instrumentation department of the IOW for operating the CTD-system and processing the CTD-data.

REFERENCES

Bernhard, J. M., Buck, K. R., Farmer, M. A. & Bowser, S. S. (2000). The Santa Barbara Basin is a symbiosis oasis. *Nature* 403, 77-80.

Bianchi, G., Carpenter, K. E., Roux, J.-P., Molloy, F. J., Boyer, D. & Boyer, H. J. (1999). *Field guide to the living marine resources of Namibia.* Rome: FAO Fisheries Department.

Bochert, R. & Zettler, M. L. (2011). Cumacea from the continental shelf of Angola and Namibia with descriptions of new species. *Zootaxa* 2978, 1-33.

Bochert, R. & Zettler, M. L. (2013). Hotspot mariner Biodiversität. Wind, Wasser und Wirbellose im Atlantischen Ozean. In: Beck E. (Ed.), Die Vielfalt des Lebens. (pp. 29-36) Weinheim, GE: Wiley-VCH.

Bodungen, B. von, John, H.-C., Lutjeharms, J. R. E., Mohrholz, V. & Veitch, J. (2008). Hydrographic and biological patterns across the Angola–Benguela Frontal Zone under undisturbed conditions. *Journal of Marine Systems* 74, 189-215.

Borchers, S. L., Schnetger, B., Böning, P. & Brumsack, H.-J. (2005). Geochemical signatures of the Namibian diatom belt: Perennial upwelling and intermittent anoxia. *Geochemistry Geophysics Geosystems* 6, Q06006, 1-20.

Cowie, G. L. & Levin, L. A. (2009). Benthic biological and biogeochemical patterns and processes across an oxygen minimum zone (Pakistan Margin, NE Arabian Sea). *Deep-Sea Research* II 56, 261-270.

Diaz, R. J. & Rosenberg, R. (1995). Marine benthic hypoxia: a review of its ecological effects and the behavioural responses of benthic macrofauna. *Oceanography and Marine Biology: An Annual Review* 33, 245-303.

Ekau, W., Auel, H., Pörtner, H. & Gilbert, D. (2010). Impacts of hypoxia on the structure and processes in the pelagic community (zooplankton, macro-invertebrates and fish). *Biogeosciences* 7, 1669-1699.

Emanuel, B. P., Bustemante, R. H., Branch, G. M., Eekhout, S. & Odendaal, F. J. (1992). A zoogeographic and functional approach to the selection of marine reserves on the west coast of South Africa. *South African Journal of Marine Science* 12, 341-354.

Emeis, K., Currie, B., Noli, K., Shidjuu, A., Bening, G., Berger, J., Brüchert, V., Endler, R., Ferdelmann, T., Finke, N., Graco, M., Haferburg, G., Heyn, T., Kiessling, A., Lage, S., Leipe, T., Mollenhauer, G., Sonnabend, K., Stregel, K., Struck, U., Treppke, U., Vogt, T. & Zemskaya, T. (2006). Meteor-Berichte 06-5, South-East Atlantic 2000. Part 2, Cruise No. 48, Leg 2: 42pp. http://www.dfg-ozean.de/fileadmin/DFG/Berichte/ Report__M48_2KS.pdf.

Emeis K.-C., Bruechert, V., Currie, B., Endler, R., Ferdelman, T. , Kiessling, A., Leipe, T., Noli-Peard, K., Struck, U. & Vogt T. (2004). Shallow gas in shelf sediments of the Namibian coastal upwelling ecosystem. *Continental Shelf Research* 24, 627-642.

Gallardo, V. A., Palma, M., Carrasco, F. D., Gutiérrez, D., Levin, L. A. & Cañete, J. I. (2004). Macrobenthic zonation caused by the oxygen minimum zone on the shelf and slope off central Chile. *Deep-Sea Res II* 51, 2475-2490.

Gooday, A. J., Levin, L. A., da Silva A. A., Bett, B. J., Cowie, G. L., Dissard, D., Gage, J. D., Hughes, D. J., Jeffreys, R., Lamont, P. A., Larkin, K. E., Murty, S. J., Schumacher, S., Whitcraft, C. & Woulds, C. (2009). Faunal responses to oxygen gradients on the Pakistan Margin: A comparison of foraminiferans, macrofauna and megafauna. *Deep-Sea Research II* 56, 488-502.

Gooday, A. J., Bett, B. J., Escobar, E., Ingole, B., Levin, L. A., Neira, C., Raman, A. V. & Sellanes, J. (2010) Habitat heterogeneity and its influence on benthic biodiversity in oxygen minimum zones. *Marine Ecology* 31, 125-147.

Gutierrez, D., Gallardo, V., Mayor, S., Neira, C., Vasquez, C., Sellanes, J., Rivas, M., Soto, A., Carrasco, F. & Baltazar, M. (2000). Effects of dissolved oxygen and fresh organic matter on the bioturbation potential of macrofauna in sublittoral sediments off Central Chile during the 1997–1998 El Nino. *Marine Ecology Progress Series* 202, 81-99.

Høgslund, S., Revsbech, N. P., Cedhagen, T., Nielsen, L. P. & Gallardo, V. A. (2008). Denitrification, nitrate turnover, and aerobic respiration by benthic foraminiferans in the oxygen minimum zone off Chile. *Journal of Experimental Marine Biology and Ecology* 359, 85-91.

Hughes, D. J., Levin, L. A., Lamont, P. A., Packer, M., Feeley, K. & Gage, J. D. (2009). Macrofaunal communities and sediment structure across the Pakistan margin Oxygen Minimum Zone, North-East Arabian Sea. *Deep-Sea Research II* 56, 434-448.

Ingole, B. S., Sautya, S., Sivadas, S., Singh, R. & Nanajkar, M. (2010). Macrofaunal community structure in the western Indian continental margin including the oxygen minimum zone. *Marine Ecology* 31, 148-166.

Jones, N. S. (1955). Cumacea of the Benguela Current. *Discovery Reports* 27, 279-291.

Jones, N. S. (1956). Cumacea from the west coast of Africa. *Atlantide Report* 4, 183-212.

Leiter, C. & Altenbach, A. V. (2010). Benthic Foraminifera from the diatomaceous mud belt off Namibia: characteristic species for severe anoxia. *Palaeontologia Electronica* 13.2.11A, 1-19.

Levin, L. A. (2003). Oxygen minimum zone benthos: adaptation and community response to hypoxia. *Oceanography and Marine Biology: an Annual Review* 41, 1-45.

Levin, L. A., Ekau, W., Gooday, A. J., Jorissen, F., Middelburg, J. J., Naqvi, S. W. A., Neira, C., Rabalais, N. N. & Zhang, J. (2009a). Effects of natural and human-induced hypoxia on coastal benthos. *Biogeosciences* 6, 2063-2098.

Levin, L. A., Gutiérrez, D., Rathburn, A., Neira, C., Sellanes, J., Muñoz, P., Gallardo, V. & Salamanca, M. (2002). Benthic processes on the Peru margin: a transect across the oxygen minimum zone during the 1997-98 El Niño. *Progress in Oceanography* 53, 1-27.

Levin, L. A., Whitcraft, C. R., Mendoza, G. F., Gonzalez, J. P. & Cowie, G. (2009b). Oxygen and organic matter thresholds for benthic faunal activity on the Pakistan margin oxygen minimum zone (700–1100m). *Deep-Sea Research II* 56, 449-471.

Michel, J., Westphal, H. & Cosel, R. von (2011). The mollusk fauna of soft sediments from the tropical, upwelling-influenced shelf of Mauritania (Northwestern Africa). *Palaios* 26, 447-460.

Mohrholz, V., Bartholomae, C. H., Plas, A.K. van der & Lass, H.-U. (2008). The seasonal variability of the northern Benguela undercurrent and its relation to the oxygen budget on the shelf. *Continental Shelf Research* 28, 424-441.

Mullins, H. T., Thompson, J. B., McDougall, K. & Vercoutere, T. L. (1985). Oxygen-minimum zone edge effects: Evidence from the central Californian coastal upwelling system. *Geology* 13, 491-494.

O'Toole, M. J. (1997). Comprehensive assessment of environmental problems and opportunities for USAID intervention in Namibia. Consultancy report, Windhoek: United States Agency for International Development.

Palma, M., Quiroga, E., Gallardo, V. A., Arntz, W., Gerdes, D., Schneider, W. & Hebbeln, D. (2005). Macrobenthic animal assemblages of the continental margin off Chile (22° to 42°S). *Journal of the Marine Biological Association* UK 85, 233-245.

Postel, L., Silva, A. J. da, Mohrholz, V. & Lass, H.-U. (2007). Zooplankton biomass variability off Angola and Namibia investigated by a lowered ADCP and net sampling. *Journal of Marine Systems* 68, 143-166.

Quiroga, E., Quiñones, R., Palma, M., Sellanes, J., Gallardo, V. A., Gerdes, D. & Rowe, G. (2005). Biomass size-spectra of macrobenthic communities in the oxygen minimum zone off Chile. *Estuarine, Coastal and Shelf Science* 62, 217-231.

Risgaard-Petersen, N., Langezaal, A. M., Ingvardsen, S., Schmid, M. C., Jetten, M. S. M., Op den Camp, H. J. M., Derksen, J. W. M., Pina-Ocho, E., Eriksson, S. P., Nielsen, L. P., Revsbech, N. P., Cedhagen, T. & Zwaan, G. J. van der (2006). Evidence for complete denitrification in a benthic foraminifer. *Nature* 443, 93-96.

Sakko, A. L. (1998). The influence of the Benguela upwelling system on Namibia's marine biodiversity. *Biodiversity and Conservation* 7, 419-433.

Sanders, H. L. (1969). Benthic marine diversity and the stability-time hypothesis. *Brookhaven Symposium in Biology* 22, 71-81.

Sarhan, T., Lafuente, J. G., Vargas, M., Vargas, J. M. & Plaza, F. (1999). Upwelling mechanisms in the northwestern Alboran Sea. *Journal of Marine Systems* 23, 317-331.

Scott, R. J., Griffith, C. L. & Robinson, T. B. (2012). Patterns of endemicity and range restriction among southern African coastal marine invertebrates. *African Journal of Marine Science* 34, 341-347.

Shannon, L. V. (1989). The physical environment. In: A. I. L. Payne & R. J. M. Crawford (Eds.) *Oceans of life off southern Africa.* (pp.11-27). Cape Town (SA): Vlaeberg Publishers.

Smith, C. R., Levin, L. A., Hoover, D. J., McMurtry, G. & Gage, J. D. (2000). Variations in bioturbation across the oxygen minimum zone in the northwest Arabian Sea. *Deep-Sea Research II* 47, 227-257.

Veitch, J. A., Florenchie, P. & Shillington, F. A. (2006). Seasonal and interannual fluctuations of the Angola–Benguela Frontal Zone (ABFZ) using 4.5 km resolution satellite imagery from 1982 to 1999. *International Journal of Remote Sensing* 27, 987-998.

Zettler, M. L., Bochert, R. & Pollehne, F. (2009). Macrozoobenthos diversity in an oxygen minimum zone (OMZ) off northern Namibia. *Marine Biology* 156, 1949-1961.

Zettler, M. L., Bochert, R. & Pollehne, F. (2013). Macrozoobenthic biodiversity patterns in the northern province of the Benguela upwelling system. *African Journal of Marine Science* 35, 283-290.

In: Upwelling   ISBN: 978-1-62948-174-6
Editors: W. E. Fischer and A. B. Green © 2013 Nova Science Publishers, Inc.

Chapter 3

UPWELLING IN THE NORTHERN BLACK SEA: DESCRIPTION, MECHANISMS AND IMPACT ON CHLOROPHYLL-A CONCENTRATION

A. B. Polonsky[*]
Marine Hydrophysical Institute, Sevastopol, Ukraine

ABSTRACT

Upwelling manifestations in the Northern Black Sea are described using long-term hydrometeorlogical routine observations and satellite data on the sea surface temperature and chlorophyll-A concentration for the warm season (May through October). It is shown that upwelling in the Northern Black Sea coastal zone is mostly due to rundown or/and Ekman mechanisms and is generated locally under certain wind conditions. During upwelling development, the sea surface cooling exceeds 5°C per 12 hours. Typical duration of an upwelling event is several days. The maximum cooling (up to 22°C) is observed in the coastal regions with narrow shelf. The typical number of upwelling events is one to four per warm season. There is also a rarer type of frontal upwelling which is generated at the edge of meandering Rim current and does not depend on the local wind. So, upwelling in the Northern Black Sea is a sporadic short-term event. However, climatic variations of regional temperature

---

[*] apolonsky5@mail.ru.

regime depend significantly on the upwelling parameters which are characterized by intense interannual-to-multidecadal variability. In general, long-term weakening of the rundown winds leads to positive trend of surface temperature in the warm season which reaches about 3°C per 50 yrs in some subregions. Upwelling events are accompanied by significant changes of chlorophyll-A concentration in the surface layer of the Northern Black Sea. Lead-lag relation between variations of the sea surface temperature and chlorophyll-A concentration varies from one sub-region to another and crucially depends on the vertical structure of chlorophyll-A concentration.

Keywords: Space-time pattern of temperature and chlorophyll-A concentration in the Northern Black Sea, interannual and longer-term variability of parameters of regional upwelling

1. INTRODUCTION:
GENERAL INFORMATION ON THE BLACK SEA AND REGIONAL UPWELLING

Black Sea (BS) is a semi-closed basin of Mediterranean type. It is situated between the Eastern Europe, Minor Asia and the Caucasus and joints the Marmora Sea (through Bosporus) and the Azov Sea (through Kerch Strait). The BS lies between approximately 40,9 and 46,6 °N. Its zonal extension is a little more than 1150 km (from about 27,5 to 41,8 °E, Fig.1).

From the geological point of view, the BS is a young saltish basin. Its relatively low salinity (around 17,5 -18 p.s.u. at the surface in the BS interior) is due to the positive fresh BS balance as a result of abundant fresh water inflow with rivers and precipitation. The BS joined the Marmora Sea and the Mediterranean basin around 7500 years ago as a result of sea level rise during the last glacial-interglacial transition. Before joining, it was a large fresh water reservoir. As a result of fast salinization after joining, associated death of fresh water biota and following formation of sharp stratification, the intermediated and deep BS waters have been contaminated by H_2S. This basin is the largest natural reservoir poisoned by H_2S in the world (Filippov,1968; Leonov, 1960; Oguz, 2008).

The bottom relief of the BS is characterized by widespread deep-sea plate in the central basin, with depth exceeding 2000 m; wide and shallow north-western shelf; and a sharp slope between the coastal region and internal deep-water sub-basin around the entire basin (except N-W part of the sea, Fig.1).

Rim current is the principal large-scale dynamic structure of the BS. It is an anticlockwise jet which is concentrated along the bottom slope. Its typical velocity is 0.5 to 1 m per sec. Rim current intensifies in winter and weakens in summer (Ivanov and Belokopytov and 2011; Oguz and Malanotte-Rizzoli, 1996; Polonsky and Shokurova, 2009).

Typical vertical thermal structure in May through October is characterized by shallow upper mixed layer (UML), sharp seasonal thermocline and cold intermediate layer (CIL) with temperature of ~6,5 to 8°C (Filippov, 1968; Oguz, 2008; Ivanov and Belokopytov, 2011; Polonsky and Popov, 2011; Fig.2). Sharp temperature stratification of the upper sea layer in summer leads to quite frequent and fast UML cooling as a result of coastal upwelling which is generated under favorable wind conditions. Drop of surface temperature during upwelling event in summer exceeds 5-10°C per 12 hours. Winter upwelling events are not accompanied by significant temperature anomalies within the UML because of its large vertical extension and absence of sharp temperature stratification in this season (Tolmazin, 1963; Polonsky and Muzyleva, 2008).

The main peculiarity of vertical stratification of the BS is determined by the inflow of relatively salty and warm water of the Marmara Sea through Bosporus into the deep layers and relatively fresh cold water at the surface. As a result, strong stable basic stratification is formed within the permanent pycnocline. The main reason for this is the difference in salinity in the surface and bottom layers (over 4 p.s.u.). At the same time, the vertical temperature stratification under CIL is unstable because temperature increases by about 1°C between the low CIL boundary and deep layers (Filippov, 1968; Ivanov and Belokopytov, 2011).

The vertical stratification described above leads to the following basic properties of the BS upwelling: firstly, the sea surface temperature (SST) hardly ever goes below 6,5-8°C even in the most intense summer upwelling; secondly, upwelling is accompanied by fast salinity increase within the UML. Upwelling-induced decrease in temperature and increase in salinity create strong thermohaline frontal zone in the coastal region. Thus, cross-frontal mixing associated with upwelling is a crucial mechanism of coastal and interior water exchange (Ginzburg et al., 2000; Polonsky and Popov, 2011).

It should be emphasized that coastal upwelling in the BS is not a permanent phenomenon. It is generated sporadically and mostly locally under certain orientation of wind vector in relation to coastline (Crepon et al., 1984; Polonsky and Muzyleva, 2012). At the same time, the favorable wind should be quite persistent to generate significant SST cooling. That is why, there are only a few such events per warm season (May through October) in the each BS sub-region, while typical duration of each event does not exceed 5 days. There is also another kind of upwelling in the BS which is generated at the edge of meandering Rim current and does not depend on the local wind. Its manifestation at the surface is usually not as strong as that of the coastal upwelling (see e.g. (Lovenkova and Polonsky, 2005; Polonsky and Muzyleva, 2012; Sur et al., 1994)).

Upwelling is accompanied by nutrient supply of the UML because of enhanced nutrient concentration below the seasonal thermocline and oxygen limitation due to decrease of O_2 concentration to the depth (Oguz, 2008). This leads to direct upwelling-induced biological consequences. One of them is dramatic change of chlorophyll-A concentration in the upper layer.

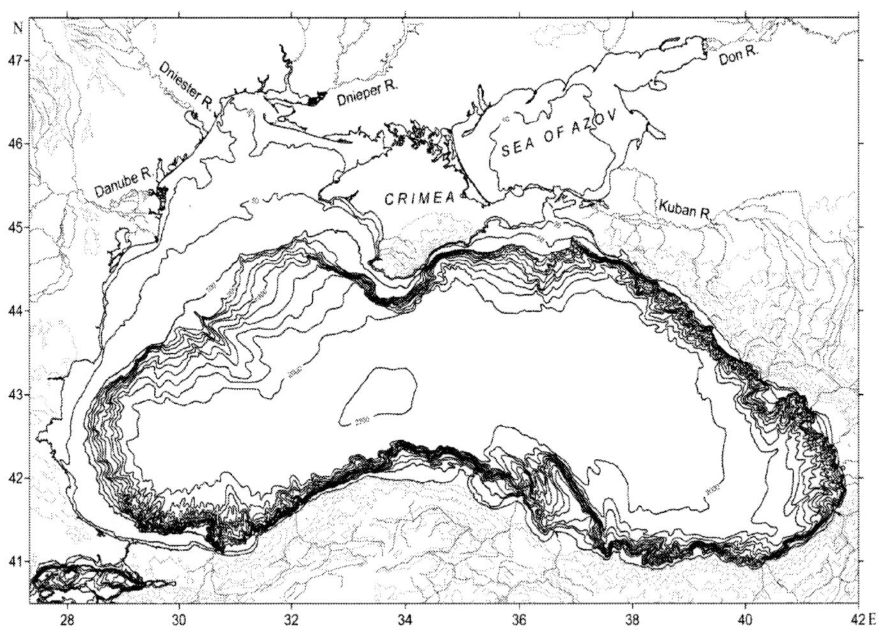

Figure 1. Map of the Black Sea including bottom relief based on the International Bathymetric Chart of the Mediterranean Sea (IBCM) (after (Ivanov and Belokopytov, 2011)).

Recent satellite data on SST and chlorophyll-A concentration permit to identify upwelling events in summer with quite high special resolution. Space-time variability of summer SST and chlorophyll-A concentration in the coastal zone of the Northern BS derived from the satellite data between 1996 and

2008 as characteristics of upwelling has been described in (Polonsky and Muzyleva, 2012). The authors showed that upwelling is accompanied by different temporal variability of chlorophyll-A in the surface layer in the N-W BS and coastal zone of the Southern Crimea.

Long-term change of regional wind regime causes variations of parameters of BS upwelling. In particular, long-term wind velocity weakening over the BS generally results in the decrease of number of the intense upwelling events in most of the coastal regions (Lovenkova and Polonsky, 2005; Polonsky and Muzyleva, 2012). However, the space-time variations of upwelling-induced long-term SST variability were not studied.

The objective of this chapter is to describe space-time structure of SST and chlorophyll-A variability in the Northern BS using more extended satellite and long-term routine observations, to discuss the governing mechanisms and to specify the relation between upwelling events and chlorophyll-A characteristics in the upper layer. It generalizes the results published by the author with different co-authors (Lovenkova and Polonsky, 2005; Mickailova et al., 2011; Polonsky and Muzyleva, 2012).

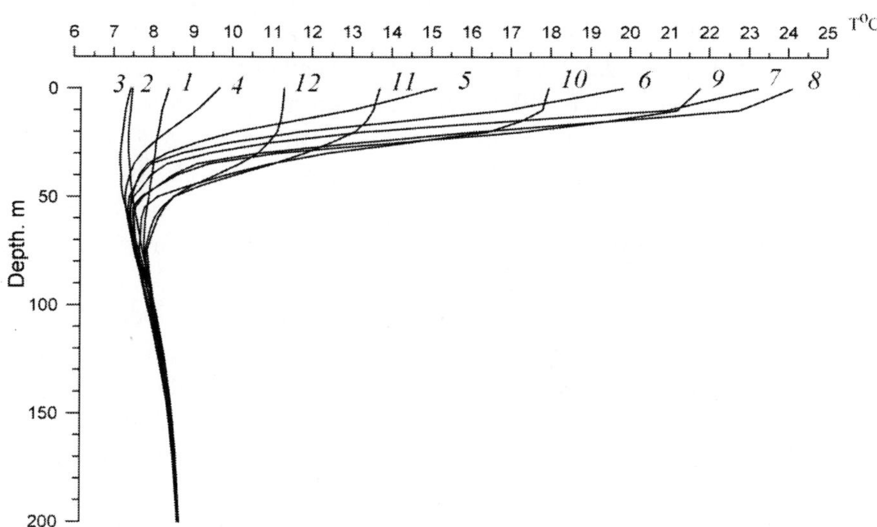

Figure 2. Mean monthly vertical profiles of temperature (°C) in the Black Sea (figures indicate the months of the year) (after (Ivanov and Belokopytov, 2011)).

2. DATA AND UPWELLING IDENTIFICATION

Satellite data for May through October of 1996-2011 collected in Department of Remote Sensing of Marine Hydrophysical Institute (http://dvs.net.ua/) were used to characterize the SST and chlorophyll-A concentration in the Northern BS. SST is registered by multi-channel radiometers (AVHHR to AVHHR-3) of NOAA satellite series, while chlorophyll-A concentration is calculated using spectrophotometer MODIS data of EOS satellite series. Typical daily number of satellite images of SST in the BS region is between 1 and 3. The larger amount of images is usually not available because of quite frequent regional cloudiness in spite of potentially higher temporal resolution (at least, for SST). Spatial resolution of the images is about 1 to 4 km.

Upwelling is identified using the following threshold: if SST decreases at least by 5°C between two sequential images (or typically 5°C per 12 hours), the upwelling event occurs. This criterion was discussed in (Lovenkova and Polonsky, 2005). Concentration of chlorophyll–A (in mg per cubic m) based on the satellite images of the BS was analyzed for each upwelling event to study the lead-lag relation between SST variations and chlorophyll-A concentration in the upper layer. It should be mentioned that satellite-based chlorophyll-A concentration differs essentially from the *in situ* observations because the algorithm of oceanic concentration evaluation should be adopted for the BS conditions (Kopelevich et al., 2002; Yunev et al., 2002; Yunev, 2009). This is why the absolute chlorophyll-A concentration is not analyzed in this chapter. I put attention just on the delay of temporal variations of the concentration in relation to the SST variability.

Long-term tendencies of winds and upwelling-induced SST manifestations were studied using routine hydrometeorological observations for May through October at the Northern BS stations in the 20th and the beginning of the 21th centuries (Fig.3). Typical interval between routine observations is 6 hrs, a few stations provide more frequent (3 hrs) measurements. Duration of available observations at the four stations (Odessa, Sevastopol, Yalta and Feodosia) exceeds 75 yrs. That is why they are analyzed in more detail. As well as for the satellite data, upwelling events were identified using selected threshold for rate of SST decrease (5°C per 12 hours).

3. GENERAL SPACE-TIME SST STRUCTURE IN THE UPWELLING ZONES OF THE NORTHERN BS AND PREDOMINANT REGIONAL UPWELLING TYPES

Monthly average SST in the N-W BS and near the Crimean coast is characterized by seasonal variability which is typical of the mid-latitudes boreal seas (Fig. 4). SST minimum occurs in January-February due to seasonal surface cooling, while SST maximum is observed in July-August as a result of summer heating. This feature is observed at all marine hydrometeorological stations and is confirmed by the numerous published results (Leonov, 1960; Ivanov and Belokopytov, 2011 and others). At the same time, the seasonal course of monthly minimum SST varies significantly from one station to another, especially in the warm period of the year (Fig. 5, Table 1). This is a result of different upwelling statistics due to various local wind conditions in the vicinity of each station. Wind roses show that the predominant wind directions in the open sea and at the regional marine hydrometeorological stations are different (Fig. 6). This is due to peculiarities of coastline and local relief.

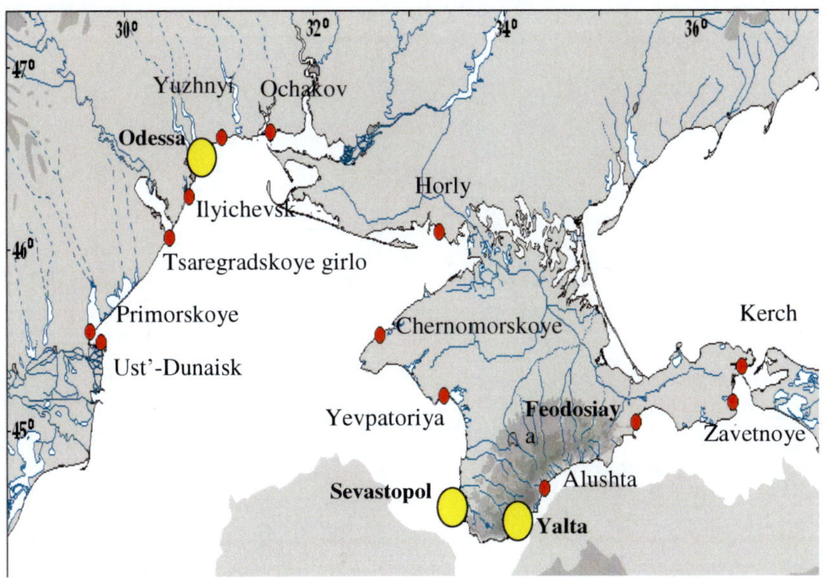

Figure 3. Location of maritime stations in the Northern Black Sea. Four stations where the length of series exceeds 75 years are shown by bold.

When I discuss the space-time SST variability in the upwelling regions of the Northern BS I use the qualification of the upwelling zones which was made in (Polonsky and Muzyleva, 2012, Figs. 7 and 8). Each zone is characterized by specific space-time SST variability due to peculiarities of bottom relief and local wind regime. Shallow N-W shelf zones are characterized by direct rundown upwelling (when wind blowing perpendicularly to the coastline causes SST cooling), while Ekman upwelling (when the wind vector coinciding with the coastline orientation generates upwelling event) is more typical of the deeper Southern Crimean coastal zone. In a few zones, where depth is close to the Ekman scale, the mixed type of upwelling (i.e., modified Ekman upwelling, when wind blowing at the angle 0 - 90° to the coastline causes sea surface cooling) is observed. Finally, region G represents the zone where the upwelling is due to a specific mechanism (Rim current meandering).

Upwelling events are very seldom generated simultaneously over all 7 sub-regions. For the period 1996 to 2011 yrs only 2 such cases were observed. They were the consequence of persistent anticyclonic blocking to the N and N-E of the BS and cyclone slowly moving along the BS periphery which resulted in strong and persistent rundown wind over the entire region (Polonsky and Muzyleva, 2012). All other upwelling events were observed in separated zones when relevant local wind conditions occurred.

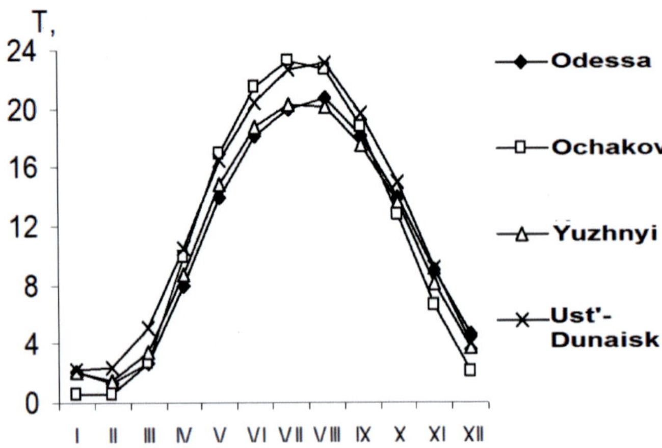

Figure 4. Variability of monthly average SST at the four stations in the N-W BS.

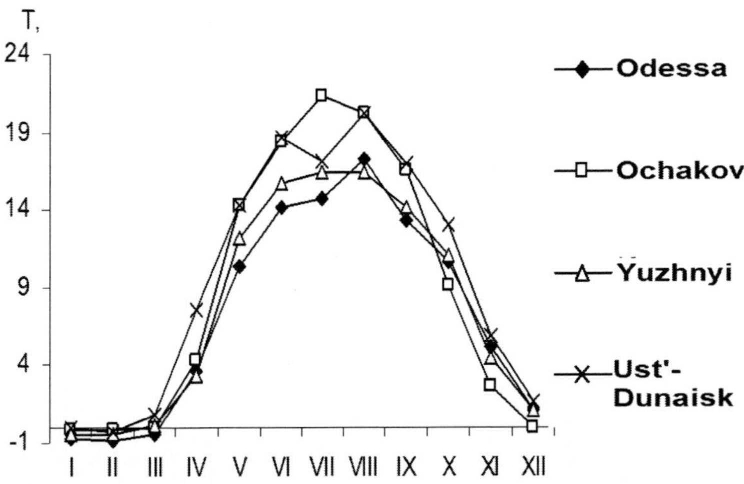

Figure 5. The same as in Figure 4, but for minimum of monthly averaged SST.

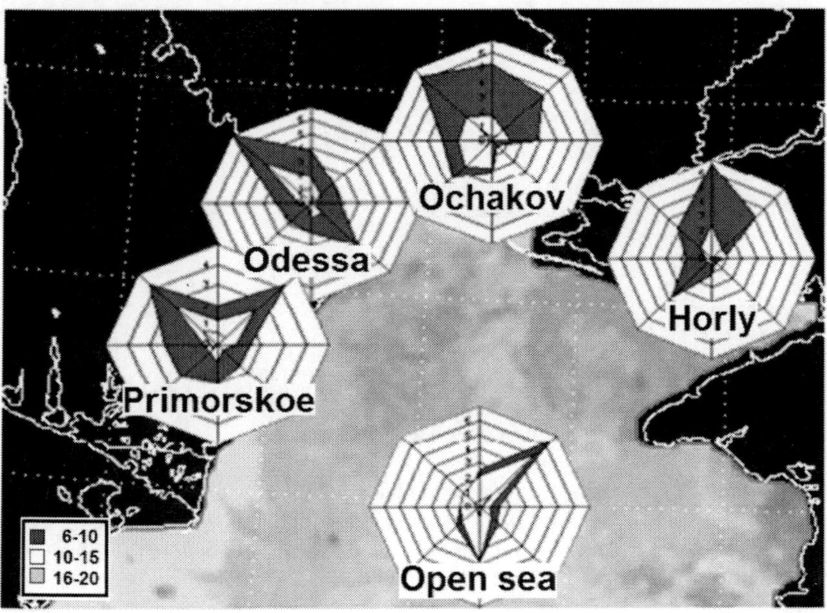

Figure 6. Climatological wind roses in the coastal zone and open interior of the N-W BS in the warm season. Routine observations (for the coastal zone) and re-analysis output (for the open interior) have been used. Wind roses for three fractions of wind speed (6 to 9, 10 to 15 and 16 to 20 m per sec) are presented.

Table 1. Characteristics of summer SST variability in Sevastopol/Yalta/Feodosia using long-term (>75 yrs) routine observations

Months	June	July	August
Monthly mean SST	20.0/18.0/17.8	23.0/21.4/20.9	23.7/23.1/22.5
Min of monthly mean	17.5/12.5/13.6	19.7/15.9/15.9	20.8/18.7/15.9
Absolute observed min	11.5/6.5/8.0	12.9/6.5/7.4	13.7/8.6/9.3
Max of monthly mean	22.9/21.6/21.4	26.2/25.4/24.8	26.8/26.1/25.9
Absolute observed max	25.5/25.5/25.5	28.2/28.5/28.4	28.1/28.1/27.7
r.m.s. of monthly means	1.2/1.8/1.8	1.3/2.0/2.1	1.2/1.3/1.7

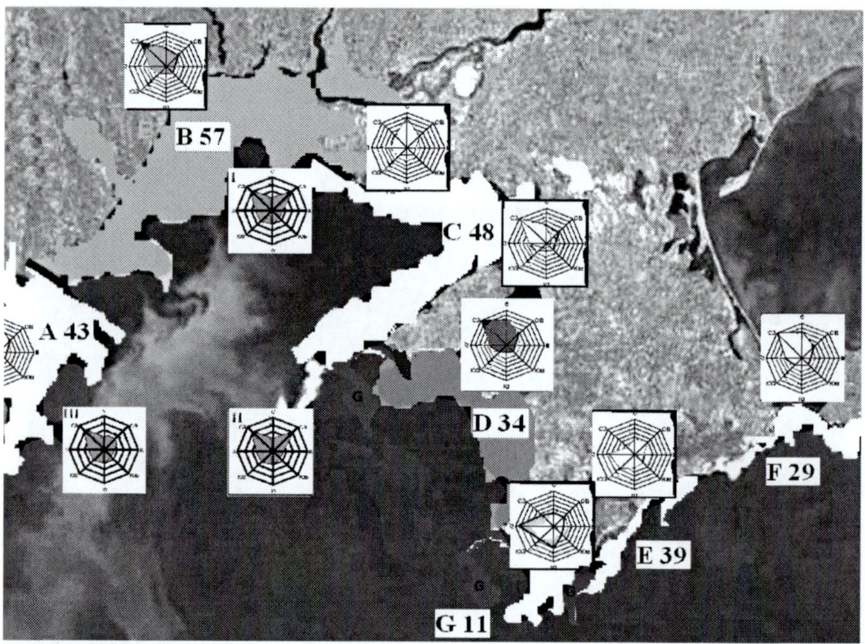

Figure 7. Total number of upwelling events identified using satellite images and routine hydrometeorological observations in May through October 1996-2011 within 7 different zones of the Northern BS. Letters A, B, C, D, E, F, G denote the zones, while figures within each zone specify the total number of the events for studied period. Predominant local wind directions during upwelling events development are done at the wind roses. Wind roses were plotted using routing hydrometeorological data and re-analysis output. The grid points of re-analysis are numbered by Roman figures.

The most numerous upwelling events were registered in the shallow Odessa region (zone B). Totally 57 events were identified using data of 1996-2011. In other words, typical annual number of upwelling events within the warm season is 3 to 4. This result is confirmed by routine SST data. In the other zones, this figure is between 2 and 3, except region G, where the upwelling events are generated even not every year (Figs. 7 and 8; Table 2).

Finally, I would like to emphasize two following things. Firstly, sea surface cooling is not always the result of upwelling. For example, zone C is usually characterized by general cooling, while for the part of this zone the wind is not rundown. As shown in (Mikhailova et al., 2011), the mechanism of SST decreasing in this zone consists of initial upwelling (with vertical velocities of about 10^{-4} m per sec) and following spreading of cold water over the entire zone due to horizontal advection. Secondly, because of the complex coastline and bottom relief of the Northern Black Sea the coastally-trapped waves do not play an important role in the upwelling dynamics there. Thus, the regional coastal upwelling can be easily identified using joint SST and wind observations.

4. INTERANNUAL VARIABILITY AND LONGER-TERM SST TENDENCIES ASSOCIATED WITH THE UPWELLING FREQUENCY CHANGE

This issue is considered using long-term observations of Ukrainian hydrometeorological service, because satellite data for the long period are not available. As a rule, general long-term wind weakening leads to less intense and frequent summer upwelling events in the second half (and especially in the last 25 yrs) of the 20 century. The number of rundown winds (and associated upwelling events) in the N-W BS has decreased between May and October by about 24% per 60 yrs (Fig. 9). This is the primary cause of multidecadal warming of the coastal waters in summer. For instance, at the following quite close Crimean stations: Sevastopol, Feodosia and Yalta, the general SST increase for May through October was 0.12, 0.11 and 2.98°C per 50 yrs, respectively, while the regional tendencies of air temperature did not coincide with SST increase (e.g., in Yalta, the summer air temperature has decreased by 1.05°C per 50 yrs). At the same time, the sea surface cooling accompanied by more frequent local rundown winds, but not accompanied by the regional monthly air cooling occurred in some other places and months. This confirms the specific upwelling origin of long-term SST decrease in the coastal zone.

Multidecadal wind weakening is not monotonous. It is accompanied by high-magnitude interannual variations of a number of rundown winds (Fig. 9).

As a result, interannual frequency of upwelling events can vary by threefold to fivefold (Table 2). Note that the number of upwelling extracted using routine hydrometeorlogical observations exceeds the same parameter counted through satellite data (cf., numbers in numerator and denumerator in Table 2) because some upwelling events could not be identified using satellite data due to tight cloudiness in the period of observations.

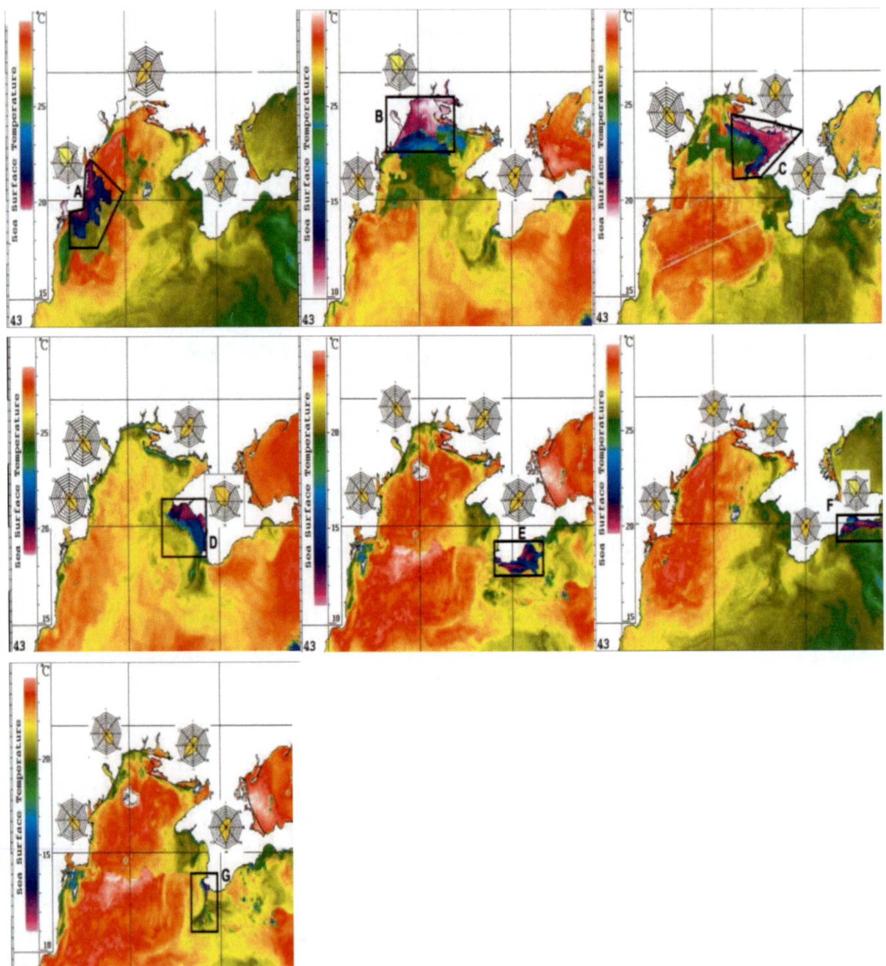

Figure 8. Selected satellite data on SST demonstrating the space upwelling-related SST patterns in 7 extracted zones. Each rectangular area denotes separated upwelling zone (A to G). Wind roses show predominate local wind directions during upwelling events.

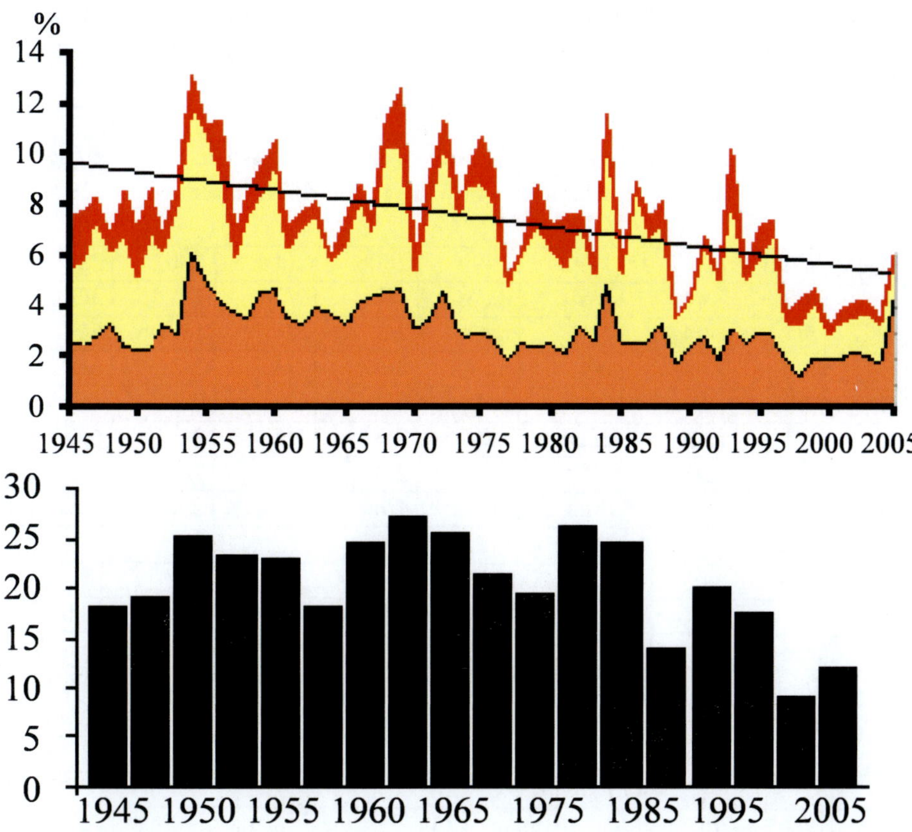

Figure 9. Interannual variability of fraction (%) of rundown winds in Odessa (upper panel) (N-W wind in red; W - in yellow and W - in brown) and total number of rundown winds from May through October there for 3 yrs intervals (low panel).

The maximum interannual difference of the summer SST occurred at the stations situated in the coastal regions with narrow shelf. For instance, in Yalta this difference in July reached 22°C. Even interannual difference of monthly mean SST extremes exceeded 9°C there (Table 1). Minimum of the summer SST temperature was observed when BS CIL temperature dropped to 6.5°C due to multidecadal regional climate variability. It happened 20 to 30 yrs ago. The recent CIL temperature is about 8.0°C (Polonsky and Popov, 2011). So, the summer SST decrease as a result of intense upwelling events cannot be below 8.0°C now.

Thus, just the change of upwelling statistics causes the significant interannual and longer-term summer SST variability in the coastal zone of the Northern BS.

Table 2. Yearly number of upwelling events in 7 specific zones of the Northern Black Sea in May through October

Zone	1996	1997	1998	1999	2000	2001	2002	2003
A	4/5	3/4	4/6	na/4	1/1	2/2	1/2	3/3
B	5/5	4/4	4/4	na/3	2/2	3/4	1/1	5/5
C	6/6	3/4	2/4	na/2	3/3	4/4	2/2	6/6
D	2/3	2/3	0/1	na/2	2/2	1/2	0/2	3/3
E	4/4	3/3	2/2	na/1	3/3	5/5	2/2	6/6
F	2/2	2/3	1/3	na/2	2/3	4/4	2/2	1/2
G	1/na	1/na	0/na	na/na	1/na	0/na	2/na	0/na

Zone	2004	2005	2006	2007	2008	2009	2010	2011	Total
A	3/3	3/3	2/2	1/1	2/2	2/2	1/1	2/2	34/43
B	3/3	5/5	3/3	2/2	4/4	2/3	3/4	5/5	51/57
C	3/3	2/2	2/2	1/1	2/2	2/2	1/2	3/3	41/48
D	2/2	2/2	1/1	3/3	1/1	1/1	1/1	5/5	26/34
E	2/2	2/2	2/2	1/1	1/1	1/1	2/2	2/2	38/39
F	1/1	1/2	0/0	1/1	2/2	0/0	1/2	0/0	20/29
G	0/na	2/na	0/na	0/na	2/na	0/na	2/na	0/na	11/n/a

Numerator – upwelling events extracted using satellite data, while denumerator -using routine observations; na – absence of data.

5. UPWELLING IMPACT ON THE VERTICAL CHLOROPHYLL-A DISTRIBUTION

All satellite images demonstrate enhanced chlorophyll-A concentration in the vicinity of the rivers' inflow because of eutrophication of these coastal waters. The vertical distribution of chlorophyll-A concentration is therefore characterized by the surface maximum there. In the subsurface layer of these zones, chlorophyll-A concentration sharply decreases. Outside of the freshened zones (those are under the impact of the rivers' inflow), chlorophyll-A concentration is at a maximum in the subsurface layer. This maximum is caused by poor eutraphication of the surface layer, relatively high concentration of oxygen and biota at the UML bottom in the inherent marine BS waters (Kapilevich et. al., 2002; Yunev, 2009).

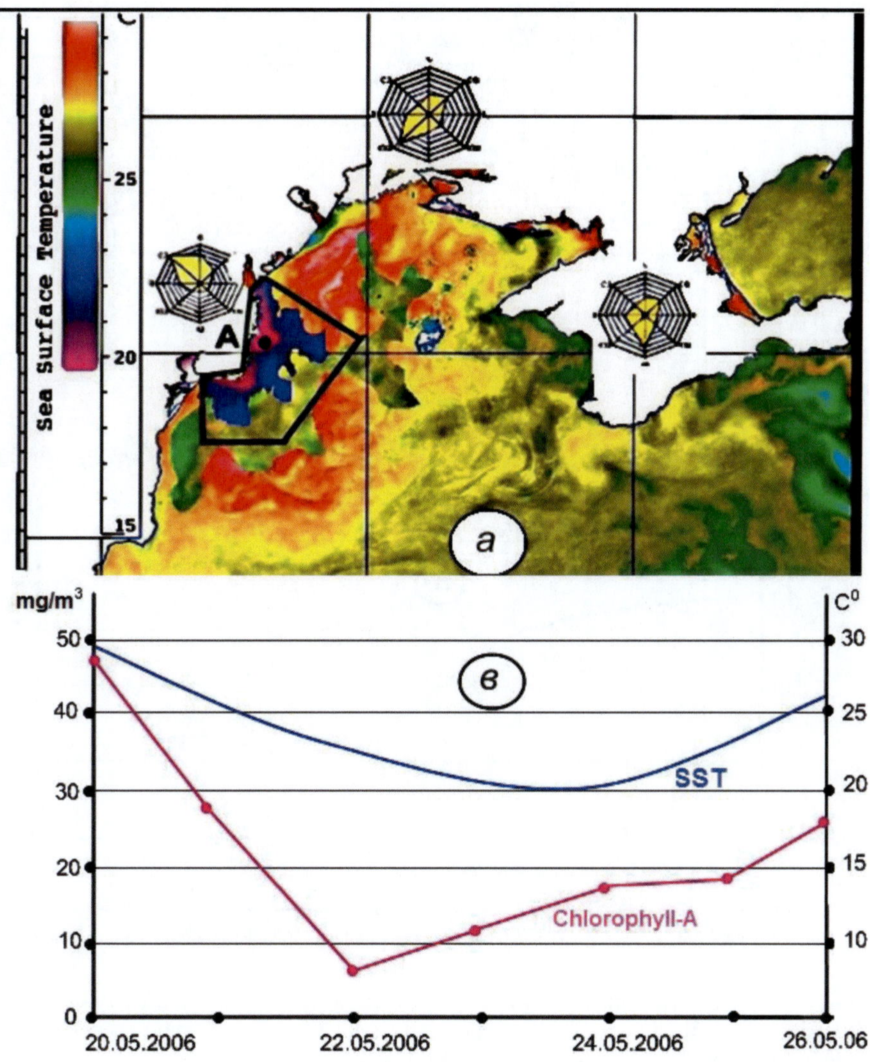

Figure 10. Space distribution of SST 22 May of 2006 (a) and time variability of SST (in blue) and chlorophyll-A concentration (in red) in the N-W BS (see black point in the zone A at the upper panel) 20-26 May of 2006 (b).

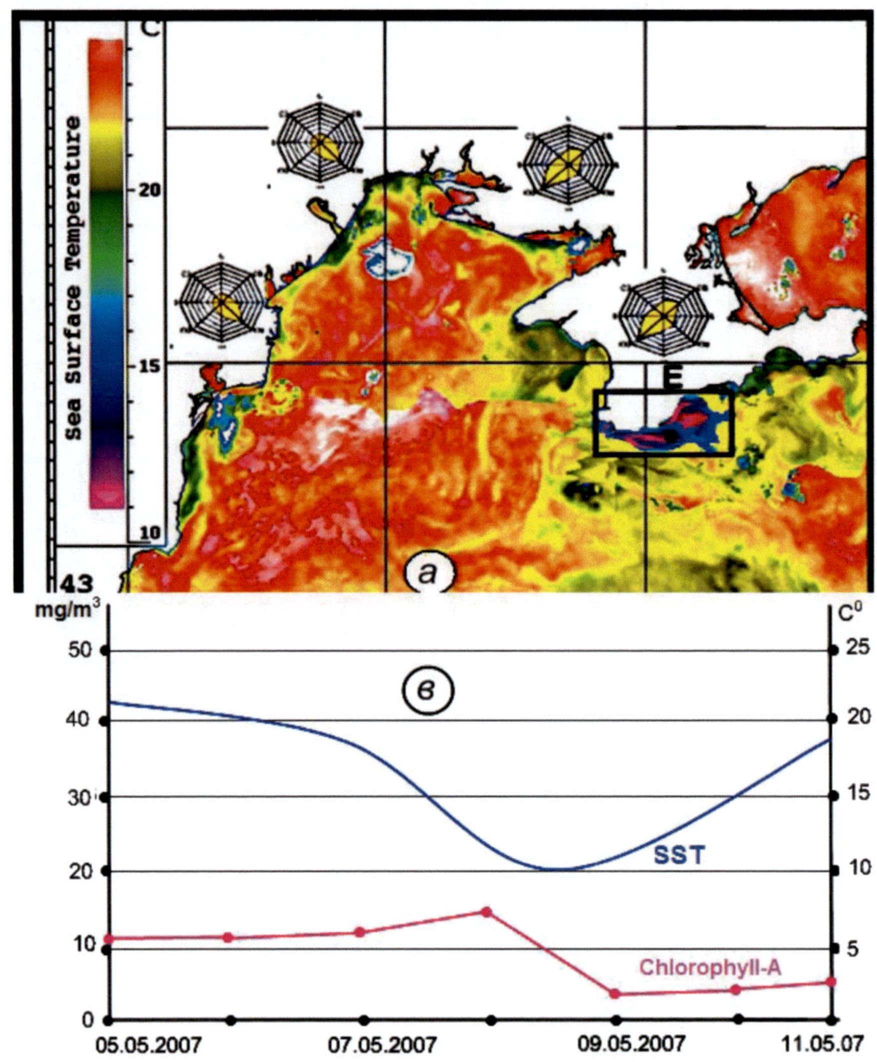

Figure 11. Space distribution of SST 8 May of 2007 (a) and time variability of SST (in blue) and chlorophyll-A concentration (in red) in the Crimean coastal region (see black point in the zone E at the upper panel) 5-11 May of 2007 (b).

Such difference of vertical chlorophyll-A distribution should lead to various lead-lag relations between peak of SST cooling under the upwelling conditions and extremes of chlorophyll-A concentration at the sea surface. At first this was shown in (Polonsky and Muzyleva 2012). Figures 10, 11 confirm this result. It is clear that chlorophyll-A concentration in the coastal zone of the N-W BS decreases when upwelling goes on, while outside of this zone the concentration first increases and then decreases. For example, during upwelling event of 20-26.05.2006 in the N-W BS, when SST dropped by about 10°C, chlorophyll-A concentration decreased by about an order. Moreover, the decrease of chlorophyll-A concentration preceded the SST drop by about 2 days. However, near the coast of the Crimea, the SST decrease during the upwelling development leads to opposite tendencies of chlorophyll-A concentration. This concentration first increases and then decreases with a lag by about 1 day after SST drop. Thus, the vertical structure of chlorophyll-A concentration is crucially important for the lead-lag relations between variations of SST in the upwelling region and chlorophyll-A concentration there.

CONCLUSION

Upwelling in the Northern Black Sea coastal zone is due mostly to rundown or/and Ekman mechanisms and is generated locally under certain wind conditions. The second (much rarer) type of upwelling is the frontal one. It is generated as a result of Rim current meandering and does not really depend on the local wind. This type of upwelling primarily impacts the coastal zone with narrow shelf.

The typical number of upwelling events in different sub-regions of the Northern Black Sea varies from 1 to 4 events per warm season (May through October). As a result of wind weakening in the second half (and especially in the last 25 yrs) of the 20th century, the number of these events has decreased by about a quarter. However, this weakening is not monotonous. There are quite strong interannual fluctuations with magnitude of the same order as amplitude of multidecadal variations or so.

Development of upwelling causes dramatic change of chlorophyll-A concentration in the upper layer. In the coastal regions those are strongly euthrophied, the maximum of chlorophyll-A concentration decrease precedes the peak of SST drop by about 2 days. The magnitude of chlorophyll-A concentration decrease reaches about an order. Outside of the euthrophied zone, upwelling event is first accompanied by chlorophyll-A concentration increase and then (with delay of about 1 day) by its decrease. This is a result of different vertical distribution of chlorophyll-A concentration in these zones.

REFERENCES

Crepon M., Richez C., Chartier M. *J. Phys. Oceanogr.*, 14, 1365 (1984).

Filippov D.M. *Circulation and Water Structure of the Black Sea*. Moscow, Nauka, 136 (1968) (in Russian).

Ginzburg A.I, Kostianoy A.G., Soloviev D.M. et al. *Satellites Oceanography and Society* (edited by D. Halpern), 273 (2000).

http://dvs.net.ua/

Ivanov V.A., Belokopytov V.N. *Oceanography of the Black Sea*. Sevastopol, EKOSEA-Hydrophysics, Sevastopol, 214 (2011) (in Russian).

Kopelevich O.V., Sheberstov S.V., Yunev O. et al. Surface chlorophyll in the Black Sea over 1978–1986 derived from satellite and *in situ* data. *J. Mar. Syst.*, 36, 145 (2002).

Leonov A.K. *Regional Oceanography*. Part 1, Leningrad, Hydrometeoizdat., 765 (1960). (in Russian).

Lovenkova E.A., Polonsky A.B. *Russian Meteorology and Hydroplogy,* No. 5, 44 (2005).

Mickailova E.N., Muzyleva M.A., and Polonsky A.B. *Physical Oceanography*, No.6, 28 (2011).

Oguz T. and Malanotte-Rizzoli P. *J. Geophys. Res.,* 101, 16,551 (1996).

Oguz T. *State of the Environment of the Black Sea (2001-2006/7),* Istanbul, Turkey, 39 (2008).

Ozsoy E. and Mikaelyan A. (Eds.). *Sensitivity to Change Black Sea, Baltic Sea, North Sea,* Kluwer Academic Publisher, Dordrecht, 510 (1997).

Polonsky A.B. and Popov Yu.I. *Recent problems of Oceanology*, MHI, Sevastopol, is.8, 54 (2011) (in Russian).

Polonsky A.B. and Shokurova I.G. *Russian Meteorology and Hydrology*, No. 4, 75 (2009).

Polonsky A.B., Muzyleva M.A. *Marine resources of the coastal zone of Ukraine*, Sevastopol, Marine Hydrophysical Institute, 257 (2012) (in Russian).

Sur H. I., Ozsoy E., Unluata U. *Progress in Oceanography*, 33, 249 (1994).

Tolmazin D.M. *Oceanology*, 3, 848 (1963) (in Russian).

Yunev O. In: *Ecological Security of Coastal and Shelf Zone and Complex Utilization of the Shelf Resources*. MHI, Sevastopol, is. 19, 252 (2009) (in Russian).

Yunev O.A., Vedernikov V.I., Basturk O. et al. *Mar. Ecol. Prog. Ser.*. 230, 11 (2002).

In: Upwelling   ISBN: 978-1-62948-174-6
Editors: W. E. Fischer and A. B. Green   © 2013 Nova Science Publishers, Inc.

Chapter 4

UPWELLING AND DOWNWELLING: DISTRIBUTION, MECHANISMS AND BIOLOGIC AND CLIMATIC SIGNIFICANCE

S. M. Govorushko

Pacific Geographical Institute, Vladivostok, Russia
Far Eastern Federal University, Vladivostok, Russia

ABSTRACT

An upwelling is a rise of deep sea waters to the surface. A minimum of four types of upwelling have been identified: (1) coastal upwelling; (2) large-scale wind-induced upwelling in the open ocean; (3) upwelling related to tropical cyclones; and (4) upwelling related to topography. The primary zones of upwelling are located off the eastern boundaries of oceans.

A downwelling is a process in the opposite direction. Basic areas of downwelling include the coastal waters of Antarctica (basically, the Weddell Sea) and the North Atlantic (predominately off the Greenland coast). The area of downwelling zones in the world's oceans is much smaller than that of upwelling zones, as the downwelling rates are always 2–3 times larger than upwelling rates.

Upwelling and downwelling are of importance for two kinds of human activity: (1) fisheries and (2) recreational activities. In addition, they play major ecological and climate-forcing roles.

1. TYPES OF UPWELLING AND THEIR MECHANISMS

An *upwelling* is a rise of deep sea waters to the surface. Accordingly, a downwelling is a process in the opposite direction. A minimum of four types of upwelling have been identified: (1) coastal upwelling; (2) large-scale wind-induced upwelling in the open ocean; (3) upwelling related to tropical cyclones; and (4) upwelling related to topography.

The emergence of coastal upwelling is possible due to external effects and internal processes taking place within the water column. An *external effect* is strong and long-lasting winds causing a negative surge, with the surface waters moving from the shore to the open ocean. A lowering of the water table near the shore occurs, and waters from the deeper horizons rise to the surface to compensate. This is the most common type of upwelling. *Internal processes* resulting in the emergence of coastal upwelling are intensification of alongshore currents and internal waves (Zalogin and Kuzminskaya 2001). The areas where coastal upwelling takes place most frequently are shown in Figure 1.

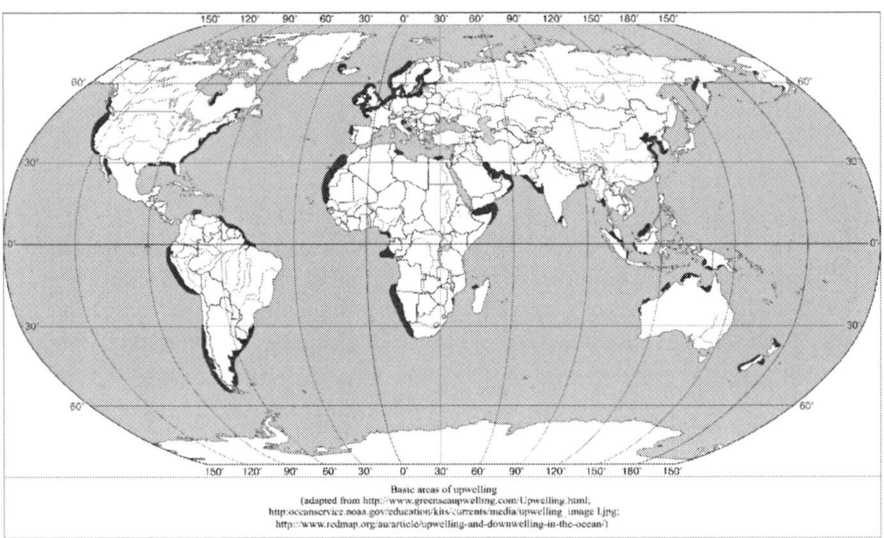

Figure 1.

Upwelling in the open ocean is observed where currents diverge (which, however, is related to the prevailing winds). When a current bifurcates and the flows diverge, the lower waters rise to the surface in order to compensate for the outgoing water. Here, the rates that water ascends are far less as compared with coastal upwelling. On the whole, upwellings in the open ocean are poorly understood.

Upwelling due to the passage of *tropical cyclones* is related to the fact that cyclone vorticity separates the water and pulls colder water from the underlying layers of the ocean to the surface. However, not all tropical cyclones result in their formation. On the average, tropical cyclones move at a speed of 20–30 km/h (Govorushko 2012), but only slow-moving cyclones (up to 8 km/h) lead to upwelling.

Upwelling related to topography emerges when islands, submerged ridges, and mounts deflect deepwater currents. Generally, they are small in area and unsteady in time. Examples include upwellings around the Galapagos Islands and the Seychelles.

2. BASIC AREAS OF UPWELLING

The *primary zones of upwelling* are located off the eastern boundaries of oceans. There are several stationary upwellings. Those in the Atlantic Ocean include the Canary (West African), Guinean, Brazilian, and South African upwellings. Those in the Indian Ocean include the Bengal and Somali upwellings. In the Pacific Ocean, there are the Chilean-Peruvian, Californian, and Oregon upwellings. There is upwelling in the Beaufort Sea (Arctic Ocean).

However, in contrast to the other upwellings, the warmer water of the Atlantic rather than cold water is supplied to the surface in the Beaufort Sea upwelling. The same upwelling is assumed to occur along the northern margins of the *Siberian* Arctic seas. The upwelling phenomenon is also characteristic of the seas. For example, it exists in summer in the middle part of the Caspian Sea off the east coast. Not infrequently, an upwelling emerges in the Black Sea (the south coast of Crimea) (Koronovsky and Yasamanov 2003). It is also observed in the Sea of Okhotsk, along the southwest coast of Kamchatka.

Many *publications* are devoted to upwelling. The results of analyzing the articles concerning the five major coastal upwellings are given in Table 1.

Table 1. Papers published on the world's main coastal upwelling areas and the major subject areas with which they deal, up to December 2011*

Subject Area/System	California	Africa	Chile	Peru	Brazil
Oceanography	883	363	261	196	76
Marine Biology	779	436	273	142	95
Geology	759	507	208	192	94
Environmental Sciences/Ecology	558	470	253	209	92
Total**	1,744	1,317	535	493	269

* Coelho-Souza et al. (2012).

** The totals given are less than the sums of the columns because for some papers, more than one area was considered.

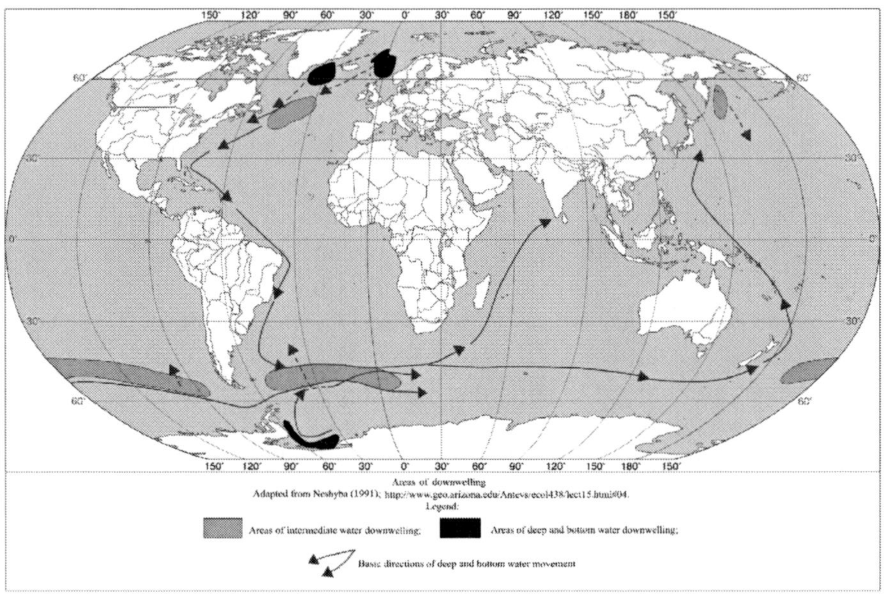

Figure 2.

3. BASIC AREAS OF DOWNWELLING

It is clear that if there are zones of upwelling, then areas of descending motion (downwelling) should exist. *Basic areas of downwelling* include the coastal waters of Antarctica (basically, the Weddell Sea) and the North

Atlantic (predominately off the Greenland coast). The area of downwelling zones in the world's oceans is much smaller than that of upwelling zones, as the downwelling rates are always 2–3 times larger than upwelling rates (Monin and Krasitsky 1985). The downwelling areas in the World Ocean are illustrated in Figure 2.

4. MECHANISMS OF DOWNWELLING

The mechanisms of downwelling differ from each other in different regions (Govorushko 2011). *In the neighborhood of Greenland,* two factors are responsible for the formation of higher-density water: (1) passage of cold, dry continental air over the ocean; and (2) high salinity of water transferred there by the Gulf Stream and its extensions (Irminger, Norway, and West Spitsbergen currents). Quick evaporation and cooling of already-dense (by virtue of high salinity) water result in increases in specific gravity and fast downwelling. The rate of the Greenland downwelling is about 5 million m^3/s (Neshyba 1991).

The densest seawater in the World Ocean takes part in the *Antarctic downwelling.* It forms in winter in the southern hemisphere. The key reason for its formation is the release of salts dissolved in the seawater when it is frozen and turns into ice. The water below the ice proves to be "overloaded" with salts to the extent that it begins to gravitate to the bottom. In the Weddell Sea, about 25 million m^3 of water with salinity of $34.68^0/_{00}$ and temperatures less than 0°C—that is, approximately 5 times more than in the Arctic downwelling—sinks to the bottom (Neshyba 1991).

The third mechanism of forming dense water is the displacement of cold and warm water masses. The water within such zones has an intermediate density; for this reason, the downwelling intensity is low. Such a mechanism exists in the zones of interaction between the Kuroshio and Oyashio currents in the Pacific Ocean and between the Labrador current and the Gulf Stream in the North Atlantic.

The fourth mechanism occurs under conditions of wind-induced surges. The water is retained against the shore, and the sea level rises. When the heavy flow of water heads landward, water is forced to descend somewhere.

**Table 2. Rates of upwelling and downwelling in different regions
of the world**

Area	Upwelling rate, cm/day	Downwelling rate, cm/day	Source
Coastal zone of California	67	–	Encyclopaedia… (1983)
Coastal zone of California	220	–	http://www.hydrometeorolog y.ws/str17.html
Nearshore zone of South America	333	–	Encyclopaedia… (1983)
Lake Baikal, south end	43–100	173–778	Shimaraev et al. (2012)
Lake Baikal, north end	8.6–560	–	Shimaraev et al. (2012)
Sea of Japan, Peter the Great Bay	3,715	–	Yurasov and Vilyanskaya (2010)
Northwestern Black Sea	3,024	–	Djiganshin et al. (2010)
New Jersey coast (USA)	800	–	http://marine.rutgers.edu/dm cs/ms501/2008/ lecturenotes/Coastal%20Up welling.doc
Southern Lake Superior	130–500	–	Niebauer et al. (1977)
Northern Lake Superior	350	–	Ragotzkie (1974)

5. QUANTITATIVE CHARACTERISTICS OF UPWELLING

According to the requirement for *mass balance* between different ocean layers and zones, upwelling and downwelling are related to each other. Water descending at one place rises to the surface in another. For example, water descending off the shores of Greenland then moves as the North Atlantic deep water, and, after hundreds of years, it rises to the surface in the southern oceans. That particular upwelling gives rise to the area's enormous productivity.

Touching on the question of upwelling and downwelling duration, one might say that these processes have been under way in the Atlantic Ocean for at least 10 million years (Dmitrenko et al. 2010). This conclusion was drawn based on analysis of core samples extracted by the drilling ships *Glomar Challenger* (40th cruise) and *JOIDES Resolution* (75th cruise).

The rates of occurrence of upwelling and downwelling in the same area can be markedly different. For example, in the southeastern Baltic Sea, upwelling and downwelling were observed, on average, for 27 and 63 days per year, respectively, during the period 1970–2010 (Golenko and Golenko 2012).

Table 3. Water surface temperature changes due to upwelling and downwelling

Area	Value of temperature fall, °C	Value of temperature rise, °C	Source
Lake Baikal	6–12		Shimaraev et al. (2012)
Caspian Sea, Dagestanian coast	7.4		Monakhova and Kuramagomedov (2012)
South-eastern Baltic Sea	4		Golenko et al. (2009)
Sea of Okhotsk, northeastern coast of Sakhalin Island	–	8–10	Rutenko et al. (2009)
Atlantic Ocean, off the coast of Namibia	8–9		Dmitrenko et al. (2010)
Bering Sea, eastern part	6		Sapozhnikov et al. (2011)
Bering Sea, western part	7		Sapozhnikov et al. (2011)
Northwestern Black Sea	4–5, up to 7		Djiganshin et al. (2010)

The rates of water rise in the upwelling area are also very different. Coastal upwellings are most intense; in this case, the rates of water rise are determined by wind speed. It is assumed that the intensity of upwelling in coastal areas is about 30 to 100 times higher than that in the open ocean (Gill 1986, V. 2). Data on coastal upwelling rates in different areas are presented in Table 2.

The range of *temperature drop* in the case of upwelling is quite broad, varying from a few degrees to 10–15°C (Djiganshin et al. 2010). Data on water temperature variations due to upwelling are given in Table 3 for different regions of the Earth.

The upwelling zone width depends on the region and factors causing the upwelling. Generally, the most intense water rise occurs within a belt 10–30 km offshore. The water rise within the upwelling zone is characterized by significant vertical deviations. In the surface layer of 10–40 m thick, horizontal motion offshore with speeds of 10–30 cm/s takes place against the background of the vertical movement. In the subsurface layer occupying the whole water column of 30–10 m above the bottom, the current moves to the shore at speeds of 2–20 cm/s. In the near-bottom layer, the current is practically absent (Zalogin and Kuzminskaya 2001). The sizes of upwelling zones in different regions are shown in Table 4.

Table 4. Sizes of upwelling zones in different regions of the world

Area	Sizes of upwelling zone, km		Source
	Minimum	Maximum	
Lake Baikal, southern	5 x 60	13 x 100	Shimaraev et al. (2012)
Lake Baikal, northern	5 x 250	13 x 250	Shimaraev et al. (2012)
Caspian Sea, Dagestanian shore	256 km^2	2,200 km^2	Monakhova and Kuramagomedov (2012)
Sea of Okhotsk, northeast coast of Sakhalin Island	–	40 x 300	Rutenko et al. (2009)
Cabo Frio upwelling system, southeastern Brazil	–	70 x 400	Coelho-Souza et al. (2012)
Peru shores	–	220,000 km^2	Nixon and Thomas (2001)

6. IMPACT OF UPWELLING ON HUMAN ACTIVITIES

Upwelling and downwelling are of importance for *two kinds of human activity* (Govorushko 2011): (1) fisheries and (2) recreational activities. In addition, they play major ecological and climate-forcing roles.

The importance of upwelling for *fisheries* is great. The deep waters are very rich in biogenic matter (phosphates and nitrates). When they rise to the surface within the illuminated zone, the productivity of the water increases sharply, and it is favourable to phytoplankton growth. In turn, this favours the growth of zooplankton and the development of all subsequent links in the food chain, including fish. Although the upwelling zones occupy a very small part of the ocean area, they provide a predominant portion of the fish that are harvested. Nevertheless, the quantitative assessments of these factors from different authors vary markedly (Table 5).

Table 5. Area of upwelling zones and share of harvesting of fish according to the data of different authors

Area of upwelling zones (% of the World Ocean area)	Harvesting of fish (% of total catch)	Source
0.1	–	Neshyba (1991)
0.1	50	Gill, V. 2 (1986)
1	20	Coelho-Souza et al. (2012)
less than 1	20	Pauly and Christensen (1995)
5	25	Jennings et al. (2001)
–	50	Blanchette et al. (2009)

The most productive sections of the ocean are upwelling zones off the shores of Peru, Bengal (India) (Heinrich and Hergt 2003), and the Canary Islands (Dukhova et al. 2009).

An upwelling has a certain effect on *recreational activities,* related to abrupt drops in the water temperature within the coastal zone. For example, short-lived wind-induced upwellings often arise in the Black Sea, decreasing the water temperature by 3°–5°C (Zalogin and Kuzminskaya 2001). In July 2004, upwelling periodically (for 5–6 days) caused the water temperature to drop at the Black Sea health resorts of the Crimea and Caucasus to 13°–14°C, with the result that the swimming season was practically ruined.

7. Environmental Importance of Upwelling

7.1. Biological Role of Upwelling

The biological role of upwelling is great. The bottom waters are rich in such biogenic elements as nitrogen and phosphorus. Their availability there is caused by destruction of submerged organic matter (generally, defunct plankton). When the bottom waters float to the surface, phytoplankton begin to actively consume the biogenes. The high values of primary productivity continue to be observable at the higher levels of the food chain (phytoplankton

→ zooplankton → filter-feeding organisms → fishes → marine mammals and birds). For example, in the vicinity of the Peruvian upwelling, the fish capacity is 100 times higher than that in the adjoining areas of the ocean. Here, about 20% of of the world fishery catch is produced (http://www. hydrometeorology. ws/str17.html).

Depending on particular conditions, the manifestations of upwelling can be diverse, but, in any case, the effect of upwelling on biological productivity is excellent. Eventually, agglomerations of organic matter at the bottom, covered with layers of deposits, can change into oil. Therefore, one can talk, to some extent, of the influence of upwelling on the *oil extracting industry*.

7.2. SIGNIFICANCE IN CLIMATE FORCING

Upwelling is of essential *importance to climate*. When temperatures at the water surface are low, the air layer contiguous to it is also cooled. Therefore, with increases in altitude, air temperatures increase rather than decrease, as is usual; that is, stable *temperature inversions* are characteristic of upwelling zones. These inversions prevent the penetration of moist marine air at higher altitudes and inland. The inversions also limit vertical cloud formation; therefore, all deposits of moisture concentrate within a thin surface layer, resulting in the frequent formation of *fogs*. Fogs are particularly noted in the nearshore zone of Sakhalin Island (Rutenko et al., 2009). In areas of intensive upwelling, precipitation is lacking during much of the year. Consequently, the narrow coastal belts in such areas are distinguished by their *arid climate*. A striking example of this is provided by the Pacific coasts of Peru and Chile. Where there is an alternation of upwelling and downwelling, water surface temperatures are, on average, lower than those in other areas (Gill, 1986, V. 2). The climatic effects of such zones are similar, although they are not particularly noticeable.

REFERENCES

Blanchette C. A.; Wieters, E. A.; Briotman, B. R.; Kinlan, B. P.; Schiel, D. R. 2009. *Trophic structure and diversity in rocky intertidal upwelling ecosystems: a comparison of community patterns across California, Chile,*

South Africa, and New Zealand. Progress in Oceanography. doi:10.1016/j.pocean.2009.07.038.

Coelho-Souza, S. A.; Lopez, M. S.; Guimaraes, J. R. D.; Coutinho, R.; Candella, R. N. 2012. Biophysical interactions in the Cabo Frio upwelling system, southeastern Brazil. *Brazilian Journal of Oceanography* 60(3), 353–365.

Djiganshin G. F.; Polonsky, A. B.; Muzyleva, M. A. 2010. Upwelling in the north-western Black Sea at the end of summer season and its causes. *Mor. gidrofiz. zhurnal.* 4, 45–57.

Dmitrenko, O. B.; Oskina, N. S.; Lukashina, N. P. 2010. Bengal upwelling in the Quaternary. *Priroda* 8, 48–51. (in Russian).

Dukhova, L. A.; Shnar, V. N.; Chernyshkov, P. P.; Smolyaninova, E. A.; Timokhin, E. N.; Remeslo, A. V.; Sapozhnikov, V. V. 2009. Study of the hydrochemical structure of the Canary upwelling waters in July–August, 2008. *Okeanologiya* 49(4), 630–633. (in Russian).

Encyclopaedia Ocean-Atmosphere. 1983. Gidrometeoizdat: Leningrad, 464 pp. (in Russian).

Gill, A. 1986. *Dynamics of Atmosphere and Ocean.* Mir: Moscow, V. 1 397 pp., V. 2 415 pp. (in Russian).

Golenko, M. N.; Golenko, N. N. 2012. On structure of dynamic fields in the south-eastern Baltic under the wind influences resulting in upwelling and downwelling. *Okeanologiya* 52(5), 654–667. (in Russian).

Golenko N. N.; Golenko, M. N.; Shchuka, S. A. 2009. Observation and simulation of upwelling in the south-eastern Baltic. *Okeanologiya* 49(1), 20–27. (in Russian).

Govorushko S. M. 2011. Upwelling and its significance for humanity. *Geographia v shkole* 7, 24–26 (in Russian).

Govorushko, S. M. 2012. *Natural Processes and Human Impacts: Interaction between Humanity and the Environment.* Springer: Dordrecht, 678 pp.

Heinrich D, Hergt M (2003) Ecology: dtv—atlas. Rybari, Moscow, 287 pp. (in Russian).

Jennings, S.; Kaiser, M. J.; Reynolds, J. D. 2001. *Marine Fisheries Ecology.* Blackwell Scientific Ltd.: Oxford, 417 pp.

Koronovsky, N. B.; Yasamanov, N. A. 2003. *Geology.* Academy: Moscow, 446 pp. (in Russian).

Monakhova, G. A.; Kuramagomedov, B. M. 2012. About the peculiarities of the upwelling at the western coast of the Middle Caspian in the summer of 2011. *Scientific journal of KGAU*, 83(09). (in Russian).

Monin, A. S.; Krasitsky, V. L. 1985. *Phenomena on the ocean surface.* *Gidrometeoizdat:* Leningrad, 375 pp. (in Russian).

Neshyba, S. 1991. *Oceanology.* Mir: Moscow, 414 pp. (in Russian).

Niebauer, H. J.; Green, T.; Ragotzkie, R. A. 1977. Coastal upwelling/downwelling cycles in southern Lake Superior. *Journal of Physical Oceanography* 7, 918–927.

Nixon, S.; Thomas, A. 2001. On the size of the Peru upwelling ecosystem. Deep sea research, Part 1. *Oceanographic Research Papers* 48(11), 2521–2528.

Pauly, D.; Christensen, V. 1995. Primary production required to sustain global fisheries. *Nature* 374, 255.

Ragotzkie, R. A. 1974. Vertical motions along the northern shore of Lake Superior. *Proc. 17th Conf. Great Lakes Res.,* 456–461.

Rutenko, A. N.; Khrapchenkov, F. F.; Sosnin, V. A. 2009. Near-shore upwelling on the Sakhalin shelf. *Russian Meteorology and Hydrology* 34(2), 93–99. (in Russian).

Sapozhnikov, V. V.; Ivanova, O. S.; Mordasova, N. V. 2011. Identification of local upwellings in the Bering Sea by hydrochemical factors. *Okeanologiya* 51(2), 258–265. (in Russian).

Shimaraev, M. N.; Troitskaya, E. S.; Blinov, V. V.; Ivanov, V. G.; Gnatovskii, R. Yu. 2012. Upwellings in Lake Baikal. *Doklady Earth Sciences* 442(2), 272–276. (in Russian).

Yurasov G. I.; Vilyanskaya, E. A. 2010. Upwelling characteristics in the Peter the Great Bay in fall–winter season of 1999–2000. *Russian Meteorology and Hydrology* 35(10), 687–694. (in Russian).

Zalogin, B. S.; Kuzminskaya, K. S. 2001. *World Ocean.* Akademiya: Moscow, 192 pp. (in Russian).

INDEX